DANGEROUS GARDEN

DANGEROUS GARDEN

The Quest for Plants to Change Our Lives

DAVID STUART

HARVARD UNIVERSITY PRESS
CAMBRIDGE, MASSACHUSETTS
2004

Copyright © Frances Lincoln Limited 2004
Text copyright © David Stuart 2004
Illustrations copyright as listed on page 208

Printed in Singapore

Dangerous Garden: The Quest for Plants to Change our Lives
was edited, designed and produced by Frances Lincoln Limited,
4 Torriano Mews, Torriano Avenue, London NW5 2RZ

Library of Congress Cataloging-in-Publication Data

Stuart, David C.
Dangerous garden: the quest for plants to change our lives/David Stuart.
p. cm
Includes bibliographical references (p.).
ISBN 0-674-01104-X (alk. paper)
Medicinal plants—History. 2. Dangerous plants—History. 3. Poisonous plants—
History. I. Title.

QK99.A1S79 2004
591.6'3—dc22 2003062492

PAGE 1 *Rauvolfia serpentina*, a tropical relative of the periwinkle (*Vinca* species)
has been used in Indian medicine for at least a thousand years, to calm mania
and other mental disturbances. Although dried material was first brought to
Europe from the Middle East by Leonhart Rauwolff in the late sixteenth century,
the plant only became used in Western medicine from 1943. This watercolour
of *c.* 1800 is by an unknown Indian artist.

PAGE 2 *Gladiolus communis* subsp. *byzantinus*, a weed of Mediterranean field margins
that is now a common and invasive garden plant, was much prized, at least until
the seventeenth century, as a means of avoiding pregnancy, an effect for which
there is no modern evidence. This lovely plate comes from Curtis's
Botanical Magazine (1805).

OPPOSITE In this watercolour of *c.* 1785 a Chinese physician is about to carry
out acupuncture on a resigned-looking patient. By then herbal medicine and
acupuncture were already exceedingly ancient and complex systems. Today there
is much debate about their effectiveness.

CONTENTS

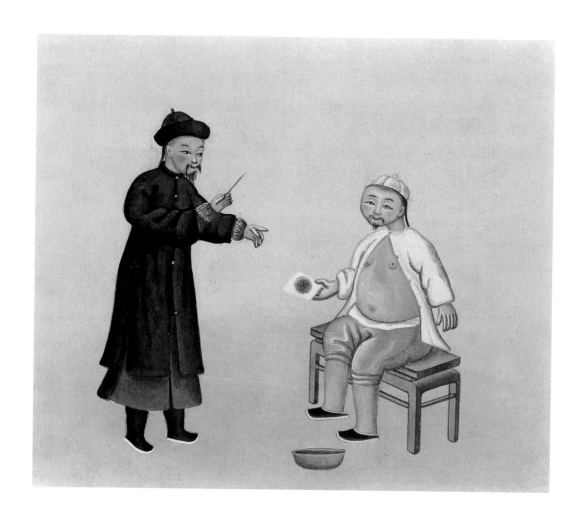

INTRODUCTION

THE IDEA FOR THIS BOOK arrived many years ago. It was a summer afternoon. I was wandering through the narrow cobbled lanes of one of those small towns that huddle for protection around the walls and gatehouse of an ancient castle. The town was pretty. Many of the gardens had apple and plum trees. There were beds of cabbages and dahlias. There were a few shops, some selling chocolates to the tourists, others bread and cakes. One of the shops, up a steep lane, I could smell from some distance away. The summer air was muddled with the sweet pear-drop smell of amyl acetate, mixed with the dusky odour of incense. The shop sold soaps in acid reds and acid yellows, Amerindian 'dream catchers', unlikely-looking crystals and sachets of herbs: chamomile, St John's wort, balm, verbena, ginseng – they were all there, prettily packaged in pastel shades, all proclaiming that they were Nature's Remedies, harvested from woods and fields, innocent, traditional, harmless, effective and devoid of side effects. It was all so sentimental, bland and often wrong. The story of the plants we have used as medicines is ancient and fascinating. It also has some very dark sides indeed.

This book is the story of how the world of plants has interacted with the human world in our search for health, happiness and longevity. It is the story of how some plants have been used as aphrodisiacs, abortifacients, intoxicants and hallucinogens; some to kill our rivals or enemies; others to deal with our gods and demons. As many of these plants have moved with us through time, and across continents, their functions have changed, multiplied or vanished altogether. Some, once so valuable that they generated wealthy empires, are now just spices on supermarket shelves. Others, once merely something to help a long trek pass swiftly, now dominate whole national economies and affect the way city-dwellers live over the entire planet.

Today, the World Health Organization estimates that about 80 per cent of the planet's population relies mainly on natural plant, or more rarely animal-derived, medicines. In industrialized countries, where modern Western scientific medicine is widely available, about 40 per cent of all pharmaceuticals are derived, at some stage, from natural sources. But in many ways, the story of the fascinating and complicated way that plants and humans interact is only just beginning. All plants are continuously evolving. When, if ever, ethnobotanists, pharmacologists and the rest have examined the entire flora of the planet they will have to start all over again. Not only will our diseases have changed, but so also will those of plants, and so will have the plants' responses to them. Plants have always had to look after themselves. Their environment is filled with hazards. They have produced viricides in an attempt to control virus attack. They have many highly effective bactericides and fungicides. Many plants contain powerful insecticides, and some of these work in extraordinary ways. Some plants contain substances that mimic insects' feeding hormones, stopping them eating. Others mimic the hormones that regulate insect growth and metamorphosis. Others still have nerve poisons, or substances that inhibit cell division in insects. Many of these insecticides can work for us as well as for the plants that contain them. For bigger pests still, some plants have substances that make grazing animals, including us, physically and unforgettably ill. It is not really surprising if some plants can mimic human hormones or interfere with the way our brains work, or our hearts beat.

Now, too, that we have power over the genes that plants contain, we can create extraordinary new plants containing almost whatever properties we choose. Pharmaceutical companies hope for great rewards. Just as, in the past, some plants have created huge riches, heroic travels, terrible wars and sometimes huge disappointments, so new discoveries hold out the possibility of huge wealth and power.

The story of the plants we have used to heal us has been shaped by early medical experiments that led inexorably onwards to the very recent birth of modern, Western, evidence-based, scientific medicine. Not all early experiments gave useful conclusions. They did, though, mark the way, both of our own history, and of the plants we have used to heal us.

One experiment, conducted some time in the late sixteenth century, was reported by the London barber-surgeon and apothecary John Gerard in his great *Herball* of 1597. The plant used was rhubarb. Though it had been a medicinal plant since Roman times at least, until the mid-sixteenth century it was only known in Europe as imported dried, and dead, roots. The Romans had thought them aphrodisiac. Medieval apothecaries sold them as a preventative for plague. Not much was known about its origins. Gerard wrote:

> *It hath happened in this as in many other forreine medicines or simples, which though they be of great and frequent use ... yet have we no certaine knowledge of the very place which produced them nor of their exact manner of growing, which have given occasion to divers to think diversely, and some have been so bold as to counterfeit figures out of their owne fancie ...*

He goes on, after denigrating other apothecaries' silly ideas: 'It is brought out of the Countrey of Sina (commonly called China), which is toward the East in the upper part of India, and that India which is without the Ganges: and not at all ... (as many do unadvisedly think) which is in Arabia the Happie ...' It does indeed come from western China, and various species are found throughout the Himalayan region. Gerard had recently got hold of some living material and was, like other apothecaries, excited. The excitement was especially strong because it was thought that the new plant might also be a cure for the newly arrived disease of syphilis.

The experiment that Gerard reports was carried out by a friend of his, John Bennett of Maidstone, Kent, a chirurgeon, on an unfortunate butcher's boy.

Being desired to cure the foresaid lad of an ague, which did grievously vex him, he promised him a medicine, and for want of one for the present ... he took out of his garden three or foure leaves of this plant of Rhubarb, which my selfe had among other simples given him, which he stamped and strained with a draught of ale, and gave it the lad in the morning to drinke: he wrought extremely downeward and upward within one houre after, and never ceased until night. In the end the strength of the boy overcame the force of the Physicke, it gave over working, and the lad had lost his agues, since which time he [Bennett] hath cured with the same medicine many of the like maladie ... By reason of which accident, that thing hath been revealed unto posteritie, which heretofore was not so much as dreamed of ...

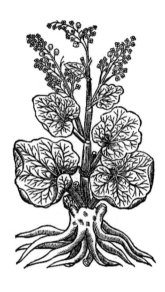

A rhubarb plant in flower – an illustration from Gerard's *Herball* of 1597. Rhubarb was once thought to cure syphilis and agues. In Central Asia, a number of species were and are used medicinally. In Europe, only its dried roots were known until living plants arrived in the sixteenth century.

Bennett clearly thought he had found a cure for ague, a form of malaria then prevalent in Britain. There were huge numbers of sufferers. Of course, rhubarb does nothing for malaria whatsoever, though, ironically, a real cure for the blood parasite that causes malaria – something which appears later in the book – had recently arrived in Europe from South America. Rhubarb isn't an aphrodisiac either, though people have hoped at some time or other that most foreign fruits and vegetables would function in this way. It certainly wasn't a cure for plague – nothing was. Nor was it a cure for syphilis. Slowly, rhubarb was realized to have no medicinal use at all, apart from its purgative nature, and there were many better plants for this. It declined slowly in favour becoming, over the next two hundred years, the pie-filling plant of late spring that we know today.

What happened to the butcher's boy illustrates a number of important aspects of the story of medicinal plants. Without controlled experiments, anecdotal evidence like Bennett's, even when multiplied a thousand-fold, does not necessarily mean that a plant is actually effective. Gerard's friend didn't check to see if the butcher's boy's fever was back the following week, as it must have been. It was widely believed at the time that a violent purge voided the substances of illness too, and so the boy, believing this, might well have felt, temporarily, that he was cured. Alternatively, he might have felt, after such a dose of rhubarb, that it was worth saying he was cured merely as a means of avoiding more medicine. Bennett didn't say. Perhaps something else was happening: the still mysterious 'placebo' effect. This phenomenon, in which a beneficial effect is produced by the patient's belief in the treatment, has been known about for a long time. Galen, doctor to Roman emperors, didactic and pompous, but an acute observer, wrote, 'He cures most in whom most are confident.' Likewise a confident and convinced patient was the easiest to cure, though Roman medicine was bizarre and often dangerous. The mind can most certainly influence the body's ability to cure itself, though it has only been since 1930 or so that the phenomenon has been much discussed and researched.

We shall see that the placebo effect manifests itself most strongly in plants with a huge 'cure list', many of which turn out, upon modern research, to have no therapeutic effect whatsoever. Rhubarb, while doctor, apothecary and patient believed in its effectiveness, may well have helped some of the patient's symptoms. Its cathartic effect, the fact that it was obviously doing something dramatic, may well have helped too. For complaints that are often 'in the mind', such as loss of libido, it may even have been a complete cure. Mayhap, even the plague victim felt a little better during the few days he had left. For the syphilis victim suffering the first symptom, it would have seemed a certain cure during the period when the disease goes underground, continuing its work but producing no symptoms at all.

The rhubarb story is also a good example of something that we shall see often in this book: a foreign and exotic plant has always elicited a more powerful placebo effect than a plant that grows in the garden next door. This is the case even if the plant has no real effect on the patient's body chemistry at all. The number of apparent cures diminished as the plant became more commonplace.

The placebo effect of plants has played a huge role in all ancient medical systems. Modern research on many frequently used Chinese, Indian, American and European medicinal plants makes it clear that many have no medicinal properties whatever. However, some do, and they have fascinating and important stories. Nevertheless, they are not numerous enough to guarantee the overall efficacy of the medical systems that used them, however ancient and exotic those traditions may be.

However, for the purposes of this book, whether a medicinal plant really works is to some degree irrelevant. The 'double blind' test, which aims to remove the placebo effect from the experiment so that neither the researcher's nor the patient's expectations affect the outcome of the result, is at the moment the best way we have of ensuring a reasonably clear-cut answer to the question: does this plant work? But it only began to be used in the 1950s and so far only a fraction of the enormous numbers of medicinal plants we have used in the past have been examined. Indeed, scientific work on the placebo effect is also only just beginning. In the past, it was irrelevant. The apothecaries and merchants in Ephesus of 200 BC thought that Chinese ephedra was a potent herb, and so they bought and sold it. Ephesus became ever more wealthy. If, in the sixteenth century, the sufferers from the new scourge of syphilis thought that dried rhubarb roots from Mongolia would cure their chancres, the merchants of sixteenth-century Venice traded in it. More marvellous palaces rose beside green canals. It is such stories that concern us here. However, in some cases, where a great deal of modern research exists for some of the more important plants, I have mentioned the plant's usefulness, or lack of it, or even its dangers, to put the plant's whole story in a fuller context.

The book is most definitely not a 'herbal'. There are no dosages. No mention of what Dioscorides or Hildegard of Bingen or John Gerard suggested for a particular disorder or difficulty should be taken as a suggestion for the plant's use today. Though many 'herbs' have little effect, some do. Any plant that can affect the metabolism of our own bodies, or of any other organism that lives on our surfaces or inhabits our interiors, should be considered extremely dangerous. Some of the plants in this book can certainly cure, but can also kill with ease.

It is estimated that about ten thousand plant species have been used in some medicinal way, in some place, or at some time in the past. All their stories are interesting, but here only a few are chosen. I have used ones that illustrate some general principle, or are of plants which are currently much in the public eye. Some simply have such wonderful stories that I wanted to re-tell them. I have also only covered flowering plants and a few ferns. Fungi I have had, alas, to ignore. I have also largely ignored the history of refined or altered plant extracts, though those of the poppy, the coca bush, fever and willow trees are just impossible to leave out. They are of huge social, political and economic importance.

Whatever effects plants may have upon us, and whatever strange things our future with the planet's plants may hold, the story, when told, is really our story. It is we who want them, whether out of vanity, desperation or fear. It is the story of the great devotion of explorers, of scientists, of people who have in their charge the care of others. It is the story of great civilizations and great cities, of business and politics. It is also the story of crime for, just as the plant world can be a dangerous place, the world of humans, with its doctors, its quacks, its dreamers and crooks, the besotted, the ignorant, the frightened and the wise, can be a dangerous place too. Perhaps most of all it is the story of our relationship to our most devoted and constant companion, Death, and our vain attempts to forget, to slow or to stop entirely our descent into his arms.

The colchicum – this one by an anonymous artist from *The Poisonous Plants of Germany and Switerzland* (1854) by C.F. Hochstetter – provides the flavouring saffron, but was once used as a mild stimulant and is sometimes still prescribed as an alleviant for gout and arthritis. However, at high doses its active substance, colchicine, causes paralysis and convulsions and is also carcinogenic.

THE GREAT
AFFLICTIONS

T HE STORY begins in a London suburb. It is the mid-twentieth century. Rank upon rank of comfortable red-roofed houses blanket the hills. 'Ena Harkness' roses fill front gardens. A few chimneys boast television aerials. Children in the street sing the old song 'Ring o' Roses', but now if someone is taken suddenly ill behind those smug front doors, a doctor is summoned or an ambulance is called. The whole weight of fast-developing Western medicine is expected to swing into action.

Yet, six hundred years ago, in the old village that now gives its name to this part of the city, when the first red blotches, the tender armpits, the sour breath, reached notice, there was no help at all.

Plague. The disease moves fast. Around groin and armpits, the lymph nodes swell rapidly into painful, pus-filled, buboes. They then deepen in colour, sometimes becoming almost black. The victim's face darkens. The sufferer may develop violent sneezing fits, and then, in a final spasm, dies. It takes four days. Only around 10 to 20 per cent of infected people survive. Cats, dogs, horses, cattle, rats, mice, all die too. Fleas abound. Suddenly death is all around. The living live in terror.

In the late 1370s, Marchione di Coppo Stefani, a Florentine aristocrat, described Europe's first brush with the new and terrible disease:

In the year of the Lord 1348 there was a very great pestilence in the city and district of Florence. It was of such a fury and so tempestuous that in houses in which it took hold previously healthy servants who took care of the ill died of the same illness. Almost none of the ill survived past the fourth day. Neither physicians nor medicines were effective ... there seemed to be no cure ... When it took hold in a house it often happened that no one remained who had not died ... Frightened people abandoned the house and fled to another. Those in town fled to villages. Physicians could not be found because they had died like the others ... Child abandoned the father, husband the wife, wife the husband, one brother the other, one sister the other. In all the city there was nothing to do but to carry the dead to a burial.

Arabic medicine carried on the ideas developed by the doctors of ancient Greece. Here, in an illustration from the 1228 Seljuk Turkish manscript of Dioscorides' *De Materia Medica*, a first-century Greek physician instructs a pupil in the use of medicinal plants.

The terrified people turned to their religion or to the hedgerow. There were huge religious processions. The price of candles became impossible. Merchants' houses, filled with sumptuous goods, were left empty and unrifled. Thieves either died or were too terrified to enter a house of death. The writer continues:

> *All the shops were shut, taverns closed; only the apothecaries and the churches remained open. If you went outside, you found almost no one. And many good and rich men were carried from home to church on a pall by four beccamorti [grave diggers] and one tonsured clerk who carried the cross. Each of them wanted a florin. This mortality enriched apothecaries, doctors, poultry vendors, beccamorti, and the grocers who sold poultices of mallow, nettles, mercury and other herbs necessary to draw off the infirmity. And it was those who made these poultices who made a lot of money.*

Yet, without any idea of what a disease was, or how it was spread, the doctors and apothecaries could do nothing to heal the sick.

Though fourteenth-century Italy was stunned by the disease's arrival, plague already had a long and virulent history. Procopius of Caesarea (modern Kayseri, in Turkey) was the first to document it, describing it in AD 541. The disease may have reached his lovely and romantic city from further east. It spread along the highway to the capital, and a year later, the Plague of Justinian began to the north-west, in Constantinople. It raged for two terrible years, killing around 70,000 of its inhabitants. People fled. Rats multiplied, then died from the plague themselves. The Byzantine emperor Justinian survived, but the economic chaos caused by the disease destroyed his attempt to re-establish his empire in the West. Later, when Muslim armies began to examine the Eastern Empire in AD 634, it was still too wrecked to resist their expansion.

The disease seems to have burnt itself out, but flared up in Europe in 1346. Though there is today some dispute that the 'Black Death' was actually bubonic plague, no other disease fits contemporary description of the disease quite as easily. Once again it seems to have come from the East. When Tartar armies had laid siege to a little town called Kaffa, on the Crimean coast, either they, or the town's inhabitants, brought the pest (another word for plague) to the fray. A few residents, already infected but escaping Kaffa, fled westwards by sea to Europe and the pest travelled with them. Italy and France were devastated by the end of 1348. The Black Death crossed the Alps to Switzerland. It reached England in August, then spread to Scotland, Ireland, Denmark, and most of Germany. By 1351, it had reached Russia. A third of the entire

OPPOSITE ABOVE Medieval Florence, walled and densely populated, made an ideal haven for the rats that carried the transmitting agent for plague – the humble flea. This 1493 woodcut is from the *Nuremberg Chronicle* by Hartmann Schedel.

OPPOSITE BELOW This plague doctor, in a seventeenth-century copper engraving by Paul Fuerst, imagined himself safe in his mask and gloves. The mask's beak contained perfumed plants such as thyme, roses and carnations, thought to filter the foul air that was believed to carry the disease.

Though the association between travel, rats and plague was made early, as shown in this fifteenth-century French painting, and tales such as *The Pied Piper of Hamelyn* were widespread, the realization that fleas were the real vector for the disease was not to come until the 1890s.

population of all those countries died, with the highest fatalities amongst children. As earlier in Byzantium, the social and economic fabric was almost destroyed.

There were recurrences of the plague for the next three centuries. It became almost commonplace. During the Great Plague of London in 1665, which killed 17,440 out of the total population of 93,000, Samuel Pepys noted almost casually in his diary: 'June 7th. This day I did in Drury Lane see two or three houses marked with a red cross upon their doors and 'Lord have mercy upon us' writ there, which was a sad sight to me.' He took it more seriously a few weeks later: 'June 21st. I found all the town almost going out of town, the coaches and wagons being all full of people going into the country.' Of course, they took their fleas with them, and so – for it was fleas that carried the disease – yet more people died.

But fleas can travel by other means. In the autumn of 1665, a box of cloth from plague-hit London reached the tiny Derbyshire village of Eyam. It was delivered to a tailor who lodged with the widow Mary Cooper in a cottage near the church. Four days later he was dead. Plague had arrived in the north. The young rector, William Mompesson, took charge. He and his Nonconformist predecessor persuaded the villagers to promise not to run away and not to allow in members of other communities. The villagers were to be fed by food left for them by people of the neighbouring village at the plague stone (which still exists). Payments were to be left in a depression in the stone filled with vinegared water. Mompesson believed that the plague was contagious from person to person. Though he was mistaken, in stopping people from meeting he inadvertently stopped the exchange of fleas. Mompesson's children happened to be already out of the village. He refused to let them return. Five villagers were dead by the end of September. Twenty-three died in October. He buried his wife, who had remained with him, in August 1666. The pest finally burnt itself out in November, by which time 260 out of 350 villagers had died. He had to write a heartbreaking letter to his children:

When plague struck a town or city, those who could fled to the countryside.
All too often, though, the travellers took infected fleas with them. Here, hopeful
London inhabitants flee the 1665 plague by carriage.

Eyam, August 31, 1666
Dear Hearts,
This brings you the doleful news of your dearest mother's death; the greatest loss that could befall you.
I am deprived of a kind and loving consort, and you are bereaved of the most indulgent mother that ever
poor little children had. But we must comfort ourselves in God, with this consideration – the loss is only
ours; our sorrow is her gain, which should sustain our drooping spirits. I assure myself that her rewards
and her joy are unutterable. Dear children, your dearest mother lived a holy life and made a comfortable
end, though by means of the sore pestilence, and she is now invested with a crown of righteousness.

Mompesson clearly thought that no apothecary and no plant could save his parish. By the 1660s, the supposed effectiveness of the old-fashioned poultices of mallow, nettles and dog's mercury was long exploded. Rhubarb, too, was known by now to be an empty hope. Few plants had arrived to take their place.

Of these, the least poisonous was valerian (*Valeriana officinalis*). The herbalist Nicholas Culpeper suggested in his *Herbal* of 1649 that 'The root boiled with liquorice, raisons and aniseed is good for those troubled with cough. Also, it is of special value against the plague, the decoction thereof being drunk and the root smelled.' If only that had been true. Those who believed Culpeper still died. Valerian is only dangerous in large doses. Perhaps, apothecaries reasoned, a cure lay in plants with a more violent effect on the human body. They turned to the yellow monkshood (*Aconitum lycoctonum*). The substances contained in this species, which John Gerard called the 'counter-poyson monkeshood', were then thought to guard humans against poisoning by the blue-flowered member of the same genus, an idea now known to be completely false: all are deadly. He also thought it 'good against the poison of all venomous beasts, the plague or pestilence and other

The common *Valeriana officinalis* of the hedgerows from *Phytographie Medicale* (1821) by Joseph Roques. Valerian was anciently used as a diuretic and cure for epilepsy. Dried roots were hung in wardrobes to perfume clothes, but – since valerian was also supposed to be overwhelmingly attractive to rats – may have encouraged their exploration by the rodents.

infectious diseases, which raise spots, pockes, or markes in the outward skin, by expelling the poison from within and defending the heart as a most sovereign cordial'. But it too saved no one, though perhaps it helped some to avoid the last difficult days of plague.

Failing that, Gerard also suggested 'mithridatum', pulverized with pimpernel water, and taken regularly. The mithridatum was a compound anti-poison supposedly first invented by Mithridates, King of Pontus (whom we shall meet again later). By the time Gerard was writing, his invention was supposed to contain around seventy-five different substances, many poisonous themselves. While some were mythical, others were only too real. Opium occupied a large percentage. Not all, even then, were convinced of its efficacy. In the play *The Knight of the Burning Pestle* by Francis Beaumont (1607), Rafe says: 'But what brave spirit could be content to sit in his shop, with a flappet of wood and a blue apron before him, selling mithridatum and dragon's water to visited houses [houses visited by the plague], that might pursue feats of arms ... ?'

Alternatively, sufferers could try a decoction of *Gentiana asclepiadea* (also dangerous in its own right), or one made of the powdered leaves of the gold-crowned blessed thistle (*Cnicus benedictus*). Sensibly, for the already ill, Gerard, sounding cautious, quotes another London doctor – 'according to the experiences of a learned gentleman M. William Godorus, Sargeant Surgeon to the Queen's Maiestie' – who claimed a cure that, administered for two or three days, 'expelleth the poyson of the pestilence, and causeth it to break forth in blisters'. The plant used was the ancient Madonna lily (*Lilium candidum*), whose sticky mucus may perhaps have softened the skin of a buboe and lessened the pain. Alternatively, devil's bit scabious (*Succisa pratensis*), or a poultice of laser wort (*Laserpitium* spp.), rue, saltpetre and honey would do the unpleasant business of bursting a buboe. But cleaning out a buboe didn't save the patient.

With everyone desperate for a cure, incautious apothecaries tried ever more poisonous things – even the newly introduced nux vomica (*Strychnos nux-vomica*) (of which far more later). Smokes from various plants were tried too, though more to save the living than to cure the ill. It was believed that the plague was spread by a cloud of poisonous gas, or a miasma, colourless but deadly, which it was hoped could be side-stepped by breathing through a strong-smelling bunch of flowers, or air filled with perfumed smoke. In southern Europe, rosemary was the most commonly burnt plant, but in northern Europe, where rosemary is not always hardy, people tried different ones. In the port city of Leith, near Edinburgh, which was sorely afflicted by plague in 1645, Alexander Abercrombie was deputed to gather heather from all around the town. This he burnt inside the affected houses. People coughed and their eyes ran, but their fleas continued biting. As heather was used as a thatching material for humble Scottish houses, perhaps Abercrombie merely raided the roofs of houses already emptied by disease. Heather smoke preserved at least him, for he survived until 1656. A London doctor, Nathaniel Hodges, developed these ideas. He encouraged those victims he came across to burn anything they could to create not only smoke but also heat; he believed that sweating out the disease was a sound approach.

But no plant could stop the plague. In western Europe, the lessening of its impact was due almost entirely to the gradual improvement of standards of civic cleanliness. Where conditions for rats remained good, the plague found new footholds. Though it reached the Americas in the sixteenth century, habitation there was sparse and it spread slowly. The French plant collector Joseph Dombey found himself in the middle of an epidemic when he arrived in the Chilean town of Concepcion in March 1782. It remained widespread in Asia into modern times. Another French plant collector and missionary priest, the Abbé Delavay, contracted the plague in China in 1886. He survived, but was partly paralysed for the rest of his life. In the United States, the last urban plague epidemic occurred in Los Angeles in 1924–5. Since then, human plague in the United States has occurred as mostly scattered cases in rural areas, usually around a dozen a year. In 2000, half a million new cases were reported in India alone. Worldwide, there are at least two million sufferers a year.

In 1894, Alexandre Yersin and Kitasato Shibasabaro, working in Tokyo, discovered the bacillus that was the infectious agent of bubonic plague. They found enormous quantities of it in the buboes of the dead. Soon after, it was shown that fleas taken from infected rats contained plague bacilli. In 1897–8, P.L. Simond proved that fleas transmitted plague. He also showed that rats that had been dead for more than a day were not dangerous to handle, as the fleas had gone. Today, only antibiotics are effective against the bacteria, now called *Yersinia pestis*. Even now, if an infected person is not treated promptly, the disease is likely to cause major illness or death.

The 'scientific' medicine upon which modern Western society relies is very new. The stethoscope was invented in 1809. The syringe was invented in the 1850s. Pasteur staged his first demonstration of anthrax vaccine in 1881, and the first vitamins were discovered in 1912 by Casimir Funk. Penicillin, discovered in 1928, was not mass-produced until 1943. Nowadays, amazing new drugs constantly appear. Before the beginning of the twentieth century all attempts at curing diseases were based almost entirely on plants.

Around 1550 BC, in Egypt, a huge papyrus was written that lists 800 medicinal drugs, including anise, caraway, cassia, coriander, fennel, cardamon, onions, garlic, thyme, mustard, sesame, fenugreek, saffron and poppy seed. Many of these plants are still in use; a number are of vast importance. Discovered in 1884 by Georg Ebers, the work is referred to throughout this book as the Ebers papyrus. In India around 500 BC, the *Sushruta-Samhita*, a herbal by a physician named Sushruta, described 700 medicinal plants, including the banana, a medicinal plant in sixteenth-century Europe. The oldest-known Chinese herbal, the Classical Pharmacopoeia of Tzu-I, was written in the same century. In eastern Europe in the fourth century BC, Hippocrates listed not only Mediterranean natives, but also exotics such as cinnamon, which were already being traded from India and the Far East. Theophrastus (c. 372–287 BC) had 550 to work with. Dioscorides listed about 650 species of medicinal plant in AD 50. Then there was a long silence, plantwise, at least in the West, until the first modern European pharmacopoeia appeared in Florence in 1498. In the East, particularly the Near East, where Greek and Roman influences survived in a way that they hadn't in Europe, medical botany flourished. It attracted men of huge and original minds like the famous Avicenna, as he was known, who influenced apothecaries into modern times. Properly called Ibn Sina, he was born in AD 980 at Afshaneh near Bukhara. His major contribution to medical science was the *Qanun fi al-Tibb*, an immense encyclopedia synthesizing medical knowledge from ancient and Muslim sources with his own acute observations. He was the first to document the contagious nature of tuberculosis, the distribution of diseases by water and soil, and the interaction between psychology and health. He listed 760 drugs, most of which were plant-based. Avicenna and many like him were, in fact, scientists. They tested things out, and their observations were acute and rational.

Yet it was in fifteenth-century Italy that the scientific idea in which contemporary observation was regarded as more useful than ancient orthodoxy took strongest root. The need to have a stable and generally accepted naming of medicinal plants fuelled the emerging science of botany. Most early botanical gardens were attached to medical schools in European universities. Medical students had to study botany. At the university of Edinburgh, founded in 1582, this practice didn't stop until the 1960s. Yet no development of sixteenth-century science helped doctors or apothecaries defeat the plague. One reason, perhaps, is that the course of the disease is so terrifyingly fast that experimentation can hardly take place before the infected victim dies. People suffering slower scourges, that allow time to try cures from the hedgerow or from distant lands, perhaps have found better succour from the plant world.

It is the year AD 1101. The queen has arrived. Surrounded by courtiers, she enters a long shadowy hall. The archbishop, also surrounded by attendants, greets her. Neither she nor he makes any cognizance of the strange pungency of the air, even though the building is almost new. Some of their attendants have their noses buried in their clothes, or in bunches of pinks and the first roses. A wooden clapper sounds and the crowd draws back. The archbishop retreats too. Queen Matilda, wife of Henry I of England and daughter of the pious Queen Margaret of Scotland, stands alone.

From the end of the room, a robed and hooded figure shuffles sideways towards her through shadows. The smell strengthens. When the apparition is close, it stops. The onlookers are fearful, fascinated and repelled. The figure raises its arms to push back its cowl and lets the upper part of its robe fall. Some of the onlookers have to turn away. It is a man. His fingers have already fallen off. His face is warped by contracted muscles, his eyelids are vastly swollen, his lower lip hangs nerveless and dripping from his mouth. His hair has mostly gone, his skin is shrinking, and he is covered with large knobbed and glistening necrotic ulcers. He suffers from unquenchable thirst.

The queen, expressionless, curtsies. She believes him to be already blessed by God. Holding out a white linen towel, she takes him in her arms and begins to kiss his sores. Many of the queen's attendants believe that she is mad. It is widely known that lepers, particularly those who are women, spread plague from sheer malevolence. Male lepers, creatures of evil too, are known to rage with sexual desire.

This strange scene takes place in one of the rooms of the leprosarium established in Canterbury in 1096. It was built to house most of the local people stricken with the disease who did not, or could not, become part of the wandering leper groups, sometimes known as *pauperes Christi*. By then there were leprosaria all over Europe, a continent in the grip of fear. Already, people were beginning to suspect that leprosy could be transmitted by contact. Sufferers were increasingly confined to institutions from AD 300 onwards. By the seventh century, there were leper houses in Verdun, Metz, Maastricht. The monastery at St Gall had one by the eighth century. The first one in Britain was set up in 996. Some historians have calculated that there may have been as many as 19,000 such institutions in

Europe. Many were dedicated to St Lazarus, and an order of knights was split off from the Knights Hospitallers to provide care for the luckless inhabitants of the leper houses.

Leprosy, nowadays politely renamed Hansen's disease, is described in ancient Indian and Chinese medical treatises. By 1400 BC Indian sacred Vedic scriptures referred to the disease as 'Kushtha', and something that may well be it is described in 600 BC by the Indian

physician Sushruta in his *Sushruta-Samhita*. It was in Japan by at least the tenth century BC, and in Egypt is mentioned in the Ebers papyrus. Galen (*c.* AD 129–99) thought that it was caught from the waters of the River Nile and helped by the unsanitary diet of the people. Egypt and India were traversed by ancient trade routes which slowly formed throughout the region to transport medicinal plants, as well as other precious goods, from around 1500 BC. The disease worked its way along them. Later, wars helped: from Egypt, it may have been disseminated around the Mediterranean by the soldiers of Gnaeus Pompey's army, returning from his campaign of 70–60 BC against Lucius Sulla in Africa. The Roman colonies of Spain, Gaul and Britain may have been infected after this date.

By the early Middle Ages, the disease was rife throughout Europe. Around the end of the fifteenth century, except for Norway and around the Mediterranean Sea, leprosy gradually eased. Isolating the sufferers had worked, though it might have been that the level of immunity in the local population had increased. In the Americas, it isn't completely clear if leprosy was present in the pre-Columbian period. That continent has a number of terrifying diseases that resemble it. However, it seems certain that the Spanish conquerors, and later the slave trade, either established or further established the disease.

The disease therefore was pitted against the plants of India and China, of Africa and Europe, and of the entire Americas. Leprosy takes a long time to kill the sufferer: death can take between eight and forty years, and sometimes it is not even fatal. So there was plenty of time for experiment with possible plant cures.

In the West, the ulcers caused by the disease were sometimes treated with the caustic juices of euphorbias, or some of the more powerfully antibacterial and antifungal plants such as alkanet (*Pentaglottis sempervirens*) and dodder (species of *Cuscuta*). Leprotic ulcers are commonly colonized by many other bacterial and fungal species and so some plants may well have helped. In China, astonishingly, frankincense was used from early times, having made its way there from the tiny kingdoms of the southern Arabian peninsula. In India, oil from spikenard or jatamansi (*Nardostachys grandiflora*) was smeared hopefully on the seeping wounds; the plant is also, and still, used as a stimulant, antiseptic and insect repellent and for the treatment of epilepsy, hysteria, convulsive affections, stomach ache, constipation and cholera. Both Ayurvedic and Unani medical systems (the latter is basically the old Graeco-Roman medicinal system of the eastern Mediterranean region) also used another ancient 'cure-all', the neem tree (*Azadirachta indica*), a plant that still plays a part in almost every aspect of Indian life.

In Europe, Dioscorides suggested treating the ulcers with darnell grass and calamintha. Galen suggested juice from a species of the dodder that parasitizes species of thyme. In late sixteenth-century London, Gerard suggested poultices of the beautiful but poisonous garden anemone (*Anemone hortensis*). New plant arrivals from the East, such as tamarind, and from South America, such as the cashew nut, were also explored as cures. Nothing worked,

Many plants were tried as cures for leprosy, and a few do reduce secondary infections of the sores caused by the disease. Here, in *Lazarus and the Rich Man, c.* 1400, by an artist of the German school, a sufferer, with his wooden warning rattle, hangs between heaven and hell.

though sometimes secondary infections were cleared up and the ulcers seemed at least slightly better.

However, lepers had some other succour. Their terrible condition, and their universal stigmatization by their local societies, gave, and gives, rise to immense mental anguish. Two plants – whose stories will be told more fully later – assuaged it. Galen suggested opium. In ancient India, the medical work by Sushruta suggested cannabis. Though of no help in saving the victims' lives, because of their effects on the brain they must have made them feel a lot better.

In the East, one plant, above all, was thought of as a cure. In Burma, now Myanmar, there was a legend that when an ancient king of that country contracted leprosy he was advised by the gods to eat the kalaw tree's fruit and was cured. In India, Hindu mythology said that when the god Rama had contracted kushtha (leprosy), he cured himself by making a medicine from the fruit of the chaulmoogra tree. It and the kalaw were the same plant. In China, too, the fruit was believed to be efficacious. Oddly, no trade in the fruit had developed, although the existing trade routes would easily have allowed the fruit's oil to travel westwards. The legends do not appear to have travelled West either.

It was not until modern medical development was in full flood in the mid-nineteenth century that the plant began to be used more widely. In 1853, the Bengal Medical Service was in the charge of Professor Dr J.F. Mouat. Unlike many of his compatriots, Mouat was interested in the potential of Indian plants as herbal medicines and listened to the legends. Treating a leprous beggar, he tried giving him crushed chaulmoogra seeds six times a day. The beggar's health improved. Mouat wrote up the experiment in *The Indian Medical Journal*. Soon, everyone wanted to try the remedy. Soon, too, natural supplies of the seeds seemed to be running short. The increasing number of leprosy cases developing in the United States meant that American pharmaceutical companies were becoming interested in the disease. So was the United States Department of Agriculture, which saw that the chaulmoogra tree might be a potentially important new crop. In 1873, Dr Armauer Hansen, a Norwegian scientist, discovered the cause of leprosy, a bacterium soon to be called *Mycobacterium leprae*, which was spreading amongst the Norwegian population and threatening to become a scourge there too. He also became interested in this mysterious tree.

The chaulmoogra is a member of the *Flacourtiaceae* family, and though it is talked of in the singular, the seeds of several closely related species produce the oil. All are usually placed in the genus *Hydnocarpus*. A typical tree produces more than 40lb of seed each season. When pressed, the kernels yield a pale yellow oil containing powerful fungicides. The remaining seed cake is sometimes used as manure, but it is too toxic to be used as animal feed. More intensive study at the University of Hawaii of oil taken from the seeds showed that, taken orally, it made patients feel extremely ill. Perhaps this hadn't mattered to Mouat's Indian

The chaulmoogra tree, a species of *Hydnocarpus* from northern India and adjoining countries, shown in a painting of 1849. Since legendary times thought of as a cure for leprosy, it turned out when investigated to have no effect on the progress of the disease.

beggar, who was at least being fed and taken care of. However, intense nausea mattered to patients in Oslo or New Orleans. The oil, which is thick and treacly, could just about be injected. The process was painful, and caused its own ulcers. Worse, it didn't seem to do much good.

Even so, another Scotsman, Dr Ernest Muir, developed a powerful belief that the oil really was a cure. Born in Scotland in 1880, he began working as a medical missionary in Bengal in 1908. By 1920, he was working in Calcutta, travelling widely in the study of leprosy. The Annual Sanitation Report of 1920 states that 'Dr Muir is engaged at the Calcutta School of Tropical Medicine in carrying out an inquiry into leprosy with an allotment of Rupees 20,000 a year of which Rupees 10,000 will be donated by the Calcutta School of Tropical Medicine.' Muir began research into chaulmoogra oil, but by 1921 he needed more seeds. The United States Department of Agriculture got to hear of him, and hatched their own plans. Finding that the chaulmoogra oil they bought from Asian markets was expensive and usually adulterated, they decided to plant hydnocarpus in Hawaii, where the climate was perfect. Also in Hawaii they discovered an extraordinary man. An egomaniac and a fantasist, he had been born in Vienna, taught himself Chinese as a youngster and, washed up on Hawaii, educated himself as a botanist. Calling himself Joseph Rock, he went on to become a famous plant collector. The trip to find chaulmoogra seeds, sponsored by USDA, was to be Rock's first expedition. Dr Muir, on meeting Rock, must have wondered quite what he was in for, as the expedition's stores included Rock's folding bath, his dining table and its accoutrements, requiring fifteen horses to carry them. Nevertheless, on 11 February 1922, Rock and Muir finally reached the far western border of China.

They eventually collected substantial quantities of chaulmoogra seed. At the end of the expedition, these were sent to interested institutions worldwide. Rock ensured that Hawaii got plenty; 27 acres of chaulmoogra trees were soon growing well at Waiahole, Oahu, and subsequently a search was made for the effective substances within the oil. But when these were at last isolated, there was a colossal disappointment. They were found to have only a dubious effect, and to be highly toxic. The King of Burma's cure, and that of Rama, existed only in legend.

In any case, chaulmoogra was soon to be overtaken. In 1908, a German chemist called Gerhardt Domack had made successful attempts to produce what became the parent chemical in the sulphone family of drugs. It was extremely toxic in humans, but effective at killing *Mycobacterium leprae* in experimental conditions. In the United States, the Parke Davis Company was successful in producing a far less toxic derivative which they named Promin. By 1940 Promin had been shown to cure rat leprosy. Three years later – after trials carried out on leprosy sufferers who were also emancipated slaves, at the end of the first year of which fifteen of twenty-two patients had improved – it was working for humans too. Chaulmoogra, and every other hoped-for plant cure, was finished.

Unsuccessful though plants were in the treatment of the plague and leprosy, some of mankind's great scourges have really found their nemesis in the plant world. Malaria is

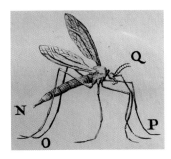 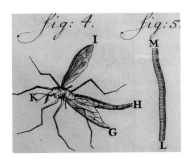 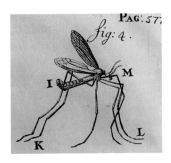

The invention of the microscope had immense impact on medicine. Here, in engravings of 1693 and 1714, A. van Leeuwenhoek shows a larva and three adults of European mosquitoes, vectors of the least dangerous form of malaria. Many plants had been tried for its cure, but only cinchona, which arrived from South America in the late sixteenth century, was effective.

another terrible disease, which killed famous men as diverse as Alexander the Great and Oliver Cromwell; it was to have immense repercussions in world politics and in global economics, and still has us deep in its thrall. It met its first true cure in an astonishing way. Unlike some of the cures tried for plague and leprosy, where the plant substances were used on the disease symptoms – the neem juices to attack secondary bacterial infections of leprotic ulcers, for instance – this plant actually gets to the core of the disease, killing the infectious agent itself.

Prehistoric man suffered from malaria in his African beginning, and the disease accompanied his migration to the Mediterranean, India and South East Asia. It also seems to have followed him up through the Bering peninsula, and then southwards down through the entire Americas. Though most important in tropical and subtropical climates, forms of it are also common in cooler regions. It was once common in northern Europe, where it was called ague. Its southern European name is the one we use today. It was once prevalent in the Pontine marshes around Rome and the Romans thought they caught it from the miasma or bad air (mal-aria) that rose from the fetid waters. Also known as Roman fever, malaria was and is generally an unpleasant and debilitating illness, but not always fatal, although some tropical types of malaria are extremely dangerous.

It wasn't until 1880 that Charles L.A. Laveran, a physician to the French colonial army in Algiers, discovered that the disease was caused by a single-celled animal, a protozoan parasite, later given the generic name *Plasmodium*. In 1902, Sir Robert Ross, an English physician in India, found that these parasites were distributed by mosquitoes. Soon after, Giovanni Battista Grassi, an Italian scientist, discovered that malaria was transmitted by the female *Anopheles* mosquito. It was not until 1948 that all the stages in its life cycle were identified.

Each of the four types of malaria is caused by its own species of *Plasmodium*. Each sort induces in the victim periodic outbreaks of intense fever. Before the early nineteenth century, most European and Asiatic plant cures were expected to reduce the victims' temperature and, by so diminishing the fever, save the life. The plants were called 'febrifuges'. While some physicians mixed their herbs, others resorted to the old Galenical system of bleeding their patients, thinking that if they could achieve a balance of blood and bile, the disease would vanish. One physician, determining that all a patient's blood was bad, tied off his arteries; unsurprisingly, the man died. Another doctor, of philosophical bent, decided that all cures were hopeless, and prescribed a reading of the *Iliad*.

While some plant febrifuges worked at reducing fever, nothing attacked the parasite. The victim remained infected and subject to periodic fevers. The disease was therefore incurable – that is, until the middle of the sixteenth century. Nicolás Monardes (1493–1588) was a medical doctor and businessman, who traded between Seville and the New World, sending cloth and slaves to America, and buying American leather and medicinal plants with the proceeds. His imports were so diverse and so prodigious that he was able to write his most important work, *La Historia Medicinal*, without leaving Seville. The work appeared in three parts (1565, 1571 and 1574), one of which contained the words:

From the New Kingdom a bark is brought, which is said to come from a large tree, which is said to have heart-shaped leaves, and bear no fruit … it has noticeable astringency to the taste, with some fragrance, because it has an aromatic aftertaste. This bark is highly regarded by the Indians, who use it against all manner of stools, with or without blood. The Spaniards, fatigued with this disease, on the advise of the Indians have used this bark and have been cured with it. They take of it as much as a small bean ground into powders, drinking them with red wine, or with proper water, as soon as they feel the fever, or malady … I received a piece of this bark two or three days ago, with which I shall experiment further, and we shall give news about all this in the third part [of this treatise] that God willing we shall write on this same matter …

The Incas seem to have known of this bark's anti-malarial virtues. In one campaign of the blood- and gold-thirsty ninth Inca emperor, Pachacutec (1438–71), his army was nearly destroyed by a terrible intermittent fever. Its remnants were saved when the bark was used as a cure. However, what had helped Pachacutec didn't vastly interest the Spanish colonists at the

Nicolás Monardes (1493–1588) was the first Western doctor to receive specimens of the fabulous 'fever bark' from Central America. He realized its importance, but did not develop it commercially. Jesuit doctors, perhaps with better American sources, did. For a while it was known as 'Jesuits' bark'. Its active substance became known much later as quinine.

time. Monardes may never have got further with his tests, or wanted to publicize the results. Nothing came of his interest in the bark. But while Don Geronimo Fernandez de Cabrera Bobadilla y Mendoza, Conde de Chinchon, was Spanish viceroy in Peru from 1629 to 1639, his wife, the Countess of Chinchon, Dona Francisca Henriquez de Rivera, became very ill with malaria. Though frequently bled by her physician, Juan del Vega, she seemed to be dying. Desperate to try anything, del Vega turned to the local Indians, and bought from them a local remedy called quina bark. After taking it, the Countess recovered immediately. Malaria was a European danger too, especially in Rome, and the bark soon arrived there. Now called the Countess's bark – Linnaeus later gave it its Latin name *Cinchona officinalis*, mistakenly leaving out one of the countess's 'h's in the genus – it was promoted by Cardinal de Lugo, the Procurator General of the Order of the Jesuits. At first, as is so common with new drugs from distant lands, it was thought to be a general cure-all, developing the big 'cure list' typical of many plants that we now know have no physiological effect. However, the cardinal supported his order's research into the proper dose of quina bark for malaria. Finding that it worked and, astonishingly, that the disease sometimes vanished entirely from the victim's body, the Jesuits soon profitably controlled the supply of quina bark to the West. Its common name now changed to 'Jesuits' bark', and it was certainly far more than a placebo.

The fact that it was named after its proponents turned out to be of great historical importance to Britain. During the Civil War of 1642–53, anti-Catholic prejudice was so strong that many Protestants, even in the depths of their fevers, refused to use it. The most celebrated objector was Oliver Cromwell, who died of malaria, or its complications, on 3 September 1658, attended by Dr Bates. Notwithstanding such a notable refusal, by 1681 the bark was universally accepted as a cure for malaria.

It was explosively successful. Wild trees were over-exploited, and soon became rare. Growing on the eastern side of the Andes at elevations from 3,000 to 9,000 feet, there turned out to be many species. Only a few had the valued curative properties. The French botanist La Condamine wrote, in 1738, of *Cinchona officinalis*, 'It is very difficult to find seeds that are ripe on the tree itself . . . they dry, & the shaking of the wind makes them fall; thus I was never able to find other than the fruit in bud on the branch, but still green as it forms just after the flower drops, or else dry & empty seed capsules.' If there was no seed in the local populations, there would be no chance of bringing the species into cultivation.

One of the first men to realize the danger that the plant might become extinct, and the advantages of cultivating it, was a Spanish doctor living in Bogotá. Born at Cadiz on 6 April 1732, José Celestino Bruno Mutis studied medicine at Seville and Madrid. He was fascinated by botany, and when he became physician-in-ordinary to the newly appointed Viceroy of New Granada, the Marqués de la Vega, he was able to study a completely new flora. Formidably energetic, he taught mathematics, founded an observatory, became canon at the cathedral of Bogotá and, of most importance, maintained two botanic gardens, at Cácota and La Montuosa. He sent plants and specimens to Linnaeus and other European botanists. The great explorers and collectors Alexander von Humboldt and Aimée Bonpland stayed at his estate for three months. It was a good place to be. Bonpland was suffering badly from malaria. They admired his plantation of the local species of *Cinchona*, only several of which turned out to contain the tree's effective substance, quinine. Mutis's plantations eventually became of immense economic importance: in 1767, the region produced no cinchona bark at all, but in 1806 it was exporting nearly a million and a quarter pounds of the precious stuff.

Though Mutis published a few papers about quina bark in South American journals, he sent his main work and his botanical specimens to Madrid. Both were lost in the confused political situation of the time. His quinine researches only reached the light of official publication in 1828, when the Spanish physician Hernández de Gregorio edited the first three parts of his work. The manuscript of Mutis's botanical work was eventually discovered by Clements R. Markham in a shed in the botanical gardens of Madrid. It too was soon published, first in Latin, then as *The Cinchona Species of New Granada*, appearing in London in 1867.

In Europe, scientific work on the *Cinchona* bark had already begun. In 1820, two young French pharmacists, Pierre-Joseph Pelletier (b. 1788) and Joseph-Bienaime Caventou (b. 1795), isolated an alkaloid from it. They selected the name quinine, derived from the native American word quinaquina, meaning bark of barks. They had already had successes, having isolated emetine from ipecac root (*Cephaelis ipecacuanha*, now *Psychotria ipecacuanha*) in 1817, and strychnine and brucine from the nux vomica (*Strychnos nux-vomica*) in 1818. Working on *Cinchona* in a cramped laboratory in the back of a Parisian apothecary shop, they found a method of separating quinine and cinchonine from the trees' barks. Having done this, they then went into the commercial production of quinine. Pelletier refused to exploit his discovery, publishing the methods of making quinine sulphate to the world.

By the 1850s, in spite of Mutis's plantations, the quina tree was getting very scarce indeed. There was huge demand for quinine in the tropical colonies of India, Africa and the East Indies, as well as in Europe itself. Clearly, the quina tree needed to be cultivated everywhere where there was a suitable climate. Mutis's work was not yet widely known in northern Europe. The Royal Botanic Gardens at Kew sent Richard Spruce, a botanist-explorer, to Ecuador to collect seeds. He found *Cinchona succirubra*, but that turned out to be a poor source of quinine. The botanic garden at Leiden in Holland sent its own botanist, who also returned with a species low in quinine. Finally, an English explorer and quina bark trader, Charles Ledger, collected from the mountains of Bolivia viable seed of the first medicinally valuable species. He sent the seeds back to his brother, who offered them for sale. The British authorities refused them as too expensive and anyway probably valueless. That was a grave mistake. Ledger offered the seeds to the Dutch in 1872. The Dutch bought, and soon had young plants in cultivation in their colony on the island of Java. Astonishingly, the bark from the saplings contained 10 to 20 per cent quinine when 3 per cent would have been adequate for treatment. This amazing new species, named *Cinchona ledgeriana*, gave the Dutch virtual control over the quinine market until the 1940s.

The plant world had a good run, but it eventually lost out. In 1944, William E. Doering and Robert B. Woodward of Columbia University synthesized quinine from coal tar, and the Dutch monopoly was finished. Although the analogue was too expensive to market at

Various species and subspecies of *Cinchona* contain varying amounts of quinine.
The first plants discovered were sterile and seedless; this illustration of *C. officinalis*
of 1774 shows seedpods as well as attractive flowers.

first, it eventually replaced natural quinine. Western science has since produced other drugs that attack *Plasmodium*: chloroquine, mefloquine and others.

Like all plant drugs that actually work, quinine, whether natural or artificial, has side effects. However, unlike many of the plants in this book, whose side effects can be devastating, those of quinine are modest. Most notable is cinchonism. This, once common in the tropics, is a mix of tinnitus, hearing loss, dizziness, nausea, uneasiness, restlessness and blurring of vision. Some of these may well have been felt by European 'colonials' who drank too much gin and tonic on the verandas of their bungalows. Quinine is bitter, and was turned into a flavouring for the aperitif, tonic water. To treat malaria, it needs to be consumed every day because it kills the parasite only at the stage when it is living in red blood cells. In earlier stages of the life cycle, when it lives in other cells, it is not affected. In any case, mosquitoes will keep biting. Thus the tonic water became a very real daily necessity for almost anyone who could afford it, and who lived in a subtropical or tropical country. Those who drunk it cut pleasantly with gin could, without too much effort, even think of the gin as healthy, for gin itself began as a medicinal drink in the eighteenth century. The juniper berries (*Juniperus communis*) that flavoured it were thought to restore weak constitutions and preserve good health.

Many other plants have been tried as anti-malarials. In China, the main one was hemp (*Cannabis sativa*); in India, it was the ever-useful neem (*Azadirachta indica*); in the Arabian peninsula, it was qat (*Catha edulis*). However, the Chinese flora has one plant, used in the West only as a flavouring, which may have long had a real anti-malarial effect. Strangely, for a culture that seems to have documented its traditions so assiduously, the knowledge was lost until a recent excavation found written references to it. The plant was *Artemisia annua*, called qinghao in the Chinese pharmacopoeia. Pharmacologists at Walter Reed Hospital in Washington, DC have found it very effective at killing plasmodium in the test tube. Though the plant is common in the West, no knowledge of its anti-malarial property appeared in our medical traditions or reached us from the East. As we shall see, we had other uses for it. India's neem tree has also been the subject of much modern research. So far, most of the results have been negative. Alcohol extractions of leaves and seeds have been successful, *in vitro*, in killing chloroquine-sensitive and chloroquine-resistant strains of the malaria parasite. The effect may not transfer to the human body; and drinking neem-leaf teas over an extended period may lead to liver damage.

Time is getting short. Malaria now infects approximately 110 million people annually, causing up to 2 million deaths; almost half of those who die are children. Some strains of the disease are even resistant to the synthetic analogues of quinine. They are not, though, resistant to the natural extract of the cinchona tree. The tree, having once saved a colonial countess, still has a huge role to play.

In contrast to the cinchona tree and its proven effect, there are hundreds of plants that exist in the shadowlands of pharmacology. Many of these plants have romantic and sometimes hugely ancient histories and are widely used today. But they have never been unequivocally proved to work. Not surprisingly, many of these 'shadow' plants are associated with what seem to be shadowy entities such as the immune system. The effects of the immune system were noted early. In 430 BC a now unidentifiable plague swept through Athens. Describing it in his *History of the Peloponnesian War*, Thucydides noted that survivors of the attack, including himself did not suffer in subsequent outbreaks of the disease. He concluded that surviving the first challenge conferred resistance to any further attack. Similar effects were noted by Chinese and Indian doctors. However, the concept of the immune system as an anatomical system, producing its own cells and its own sites of action, did not begin to appear until the 1930s. It was not until the 1940s, and the introduction of the electron microscope, that scientists began to unravel how the immune system worked. Research continues aggressively, spurred on by a new and subtle plague: human immunodeficiency virus, or HIV.

By the 1980s, HIV was decimating the gay male populations of great cities like San Francisco and New York. By the 1990s, it was decimating the straight populations of black Africa and continues to do so. It is now in full attack upon Asia, Russia, the Balkans and South America. At first, to be infected was a sentence to an extremely unpleasant and lingering death. The huge apparatus of Western medicine swung into action. After various false starts and terrible disappointments, astonishingly, by the 1990s, Western science had, for those who could afford it, slowed the progress of the virus. For some, its effects have almost been negated. Though there is a punishing regime of pill-taking, and there are sometimes major side effects, the spectre of death may be put temporarily aside. Yet, in the West, some sufferers became deeply suspicious of Western science and some parts of its agenda. In poor countries, sufferers and their governments, not all of whom would even accept that HIV exists, could not begin to afford treatment that cost a year's wages for a few days' pills. Not surprisingly, then, the world's flora was, and still is being, ransacked for plants that might, cheaply and 'naturally', halt the virus's effects.

As the immune system itself was such a recent discovery, none of the ancient pharmacopoeias had much to offer. Some herbalists began selling what they claimed were sixteenth-century 'plague' cures. The juice of *Aloe vera*, often touted as a cure-all, was advocated for a while, generated some pyramid-selling schemes, but helped no one. Some felt that the rainforests of South America, having given us cinchona, might give us a cure for HIV. A beautiful timber tree called the *pau d'arco* (several species of the genus *Tabebuia*) was hailed as an immune booster. Little proper research has been done on it, but it was an ancient cure-all widely used by Amerindians. Demand outstripped supply. Many samples reaching sufferers turned out to be made up of mahogany slivers from sawmill floors. Though the *pau d'arco* tree has powerful bactericides and fungicides, such research as there is suggests that it has no effect whatever on the human immune system.

One plant, more than any other, perhaps offered some useful function. The purple coneflower belongs to a genus of mostly handsome herbaceous perennials, all from the grasslands of North America. Native Americans had long used the fresh juice from the roots of one of them as a cure for snakebites, wounds and various other ailments. It was taken up by the European settlers, who also widely believed it to be effective. Then, in the nineteenth century, it was turned into a miracle-working all-American wonder cure: 'Meyer's Blood

Purifier'. The contents of the bottle had at least some juices from the purple coneflower. The species that the North American Indians had used was *Echinacea angustifolia*. Meyer's medicine was in such demand that sources of this plant quickly dried up. Another species, *E. pallida*, was brought into use as well. Its chemical make-up is slightly different, but that was a minor problem. What was much more important was whether the supposedly effective ingredients of the native Americans' fresh juice would still be effective once they had been processed by the bottled medicine trade. Meyer was selling a 'tincture', an alcohol extraction, of echinacea. Effective or not, it soon had official blessing. By 1916, the roots of *E. angustifolia* and *E. pallida* were put on the National Formulary of the United States. They remained on the list for a few decades, but were eventually discredited, and removed in 1950.

However, the cure had by then taken vigorous root in Europe. Such was the demand during the 1930s, especially in Germany, that seed of *Echinacea angustifolia* and *E. pallida* could not be obtained. The herbaceous border was raided and the plummy-pink-flowered *E. purpurea* was used instead. It subsequently achieved equal status with *E. angustifolia*. *E. purpurea* was extensively researched in Germany in the 1950s, just at the time of its removal from official listing in America. Many people swear by its effects, and there is a huge quantity of anecdotal evidence that it helps cure people, at least of the common cold. It was heavily promoted as an immune booster to people with HIV. However, there is as yet no seriously consistent evidence that echinacea has any influence on the human immune system at all. While some studies have suggested that echinacea can increase certain cells in the immune system (known as T-cells), the mechanism for this, if correct, is not known. Nor is it known if increasing T-cells in this way actually augments the body's ability to fight infections. Indeed in some autoimmune diseases, a T-cell increase would not be helpful. In cases of HIV, it may even serve to boost the level of virus in the body. Other studies have suggested that long-term treatment with echinacea may actually depress the immune system.

There is much confusion among the hundreds of experiments and in the way it is used. The various species of *Echinacea* have rather different chemical make-ups. Most early chemical analyses were done with *E. angustifolia*. Most recent ones have been on *E. purpurea*. Most research, following native American usage, has been done on juices extracted from living plants. Little has been done on the sort of material that passes through the herbal medicines trade to reach the final user. Standardization of the various commercial extracts is poor, and varies from country to country. Some standards are based on the contained percentage of echinacoside, which is not now itself considered an active constituent of echinacea. Commercial preparations are often water extractions, whereas the sought-after chemicals are only alcohol soluble. Many preparations also come in capsules. One study suggested that the active site of absorption is the mouth, where the lymph nodes are stimulated to produce antibodies. Capsules neatly bypass these by reaching the stomach unchallenged. None of this stops some sellers promoting echinacea as being good for the immune system. One recently advertised the plant as:

> ... *one of the strongest immune stimulators and enhancers known. It will increase the amount of T-cells and Macrophages in your bloodstream; it can double and triple them in just a few days. It also*

increases the amount of Interferon, Interleukin, Immunoglobulin and other important natural immune chemicals present in your blood … It also initiates and speeds up recovery from chronic and long-term immune depression illnesses, diseases and degeneration.

Not surprisingly, the market for echinacea is huge. In Europe alone over-the-counter sales were equivalent to $2.4 billion in 1990, 65 per cent of which were in Germany. Not surprisingly either, many American wild populations of the relevant plants have been wiped out. Even related species are under threat. *Echinacea laevigata*, never used medicinally by Amerindians, has been reduced to sixty populations. Over two-thirds of the historic populations have been eliminated, and though some of the destruction has been the result of residential and industrial development, much has been caused by collectors. Had echinacea's effect been shown, clearly, to have benefit for immune systems, the effects would have been even more extreme.

There are many hundreds of shadow plants such as echinacea: plants for which there are accumulated piles of testimonies to cures, hordes of eager merchants, hundreds of leaflets and web pages, and sometimes large numbers of university department reports giving negative, positive or inconclusive results. There are hundreds of dire afflictions which, in spite of searches carried out by humans over millennia, and over the whole globe, have never found real cures amongst the plant world: dengue and yellow fevers, filariasis, tuberculosis, cholera, typhoid, measles, diphtheria, hepatitis, smallpox and hundreds more.

However, when we begin to look at plants that affect our bodies and its organs, rather than infectious agents that attack us, the situation is very different. There are large numbers of plants that have had huge influence on how we live. As always, there are some plants that have only been benevolent. There are many more that kill us with as much ease as they sometimes cure life's afflictions.

This enthusiastically purple form of an *Echinacea* species is one of many that have been brought into use by modern herbalists because the species originally used by native Americans has been exploited almost to extinction.

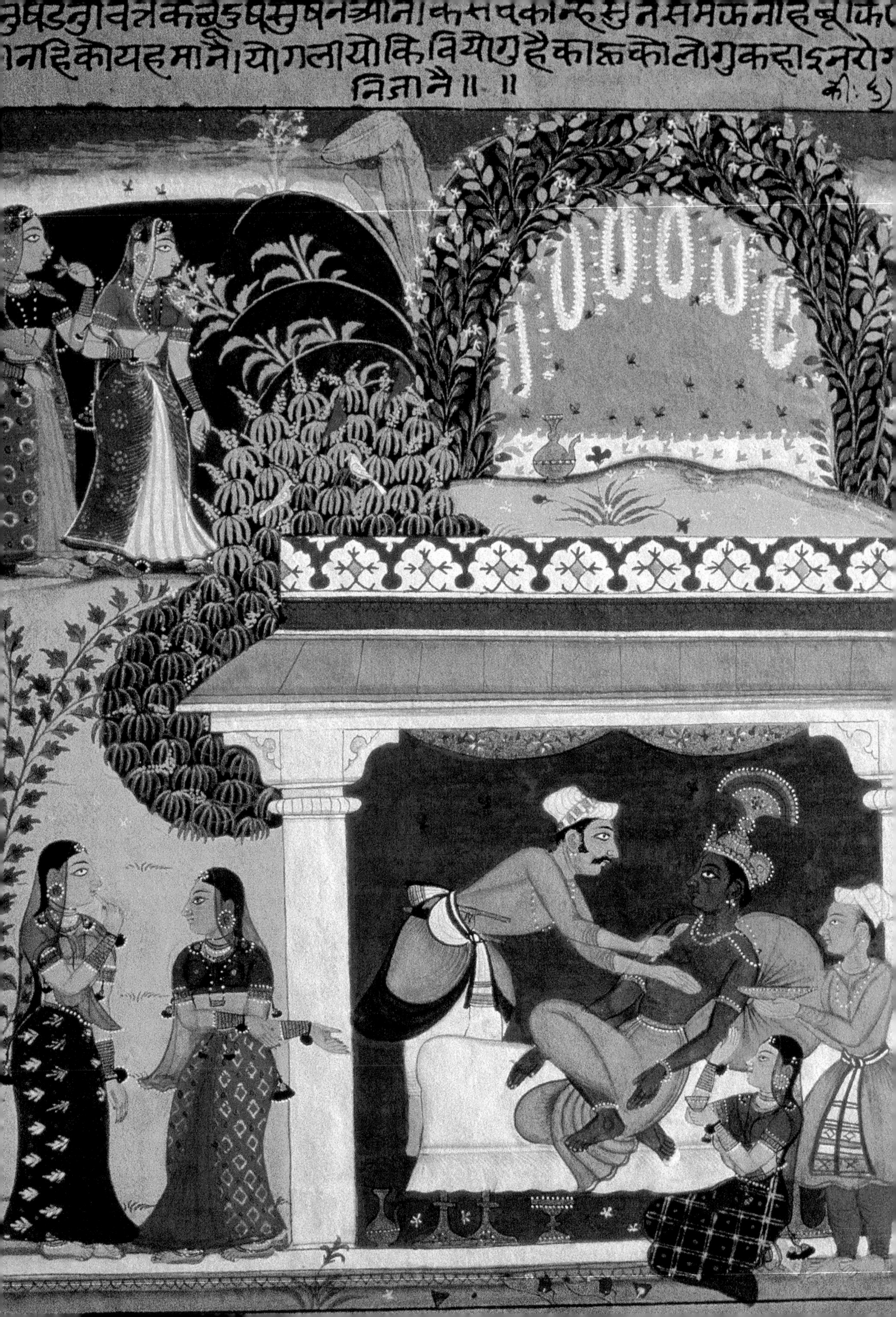
निजानै ॥ ॥

THE VITAL
ORGANS

HUMANKIND has always seen that the human body is made up of separate functional units: lose eyes and we cannot see, lose tongue and we cannot taste or speak. After every hunt, battle or accident, we see that heads and torsos also contain various other well-defined units. Until the last five hundred years or so, however, their functions in our bodies were largely unknown. We have also always theorized about 'illness', with its invisible causes and uncertain outcome. In China, since ancient times illness has been thought of as an imbalance between opposites – yin and yang, cold and heat, and so on; sometimes of particular organs, sometimes of entirely invisible aspects of the body. In India, too, illness is seen as imbalance, but as one between an individual and the environment. The closest the Indian theory of medicine comes to considering body organs is in its interest in the seven body tissues: plasma, blood, muscle, fat, bone, marrow and reproductive tissue.

In the West, theorists took in some of these Eastern ideas, and used concepts such as the 'humours' well into the seventeenth century. In the theory of the humours, imbalances of blood, phlegm, and black and yellow biles were thought to cause illness. These had correspondence to the four elements, and gave rise to four personality types. These ideas were easily integrated with the ones continuing to develop in the East. However, in the West there was also a deep underlying curiosity about what the strange organs inside the human body, iridescent and inscrutable, were for. The scientifically minded wanted to know how they were all connected. The early Greeks carried out some anatomical research on animals, but not humans. Unable to challenge the Greek taboo on defiling the human body, the great Aristotle dissected dogs, in an attempt to understand how we ourselves worked. His methods were inevitably crude. He failed to appreciate the difference between veins and arteries, though he did recognize the importance of the heart and blood. Hippocrates (460–377 BC), although looking eastwards for some of his ideas, wrote: 'In medicine one must pay attention not to plausible theorizing but to experience and reason together.' While he urged doctors to consider the influence of diet, water quality, climate and social

Even the gods get ill. Here, in a seventeenth-century illustration to the *Rasikapriya*, a lovesick deity worries about the state of his heart, seen by many cultures as the seat of emotion.

37

environment on illness much in the Ayurvedic manner, his advice spurred on those eager to find out how the body worked.

Blood, being the first thing that appears at any wound, fascinated early scientists, and was particularly important in the writings of Galen. Exploring beyond blood, he built up a rather rough guide to the interior of the mammalian body and had a huge influence, not always benign, on the development of European medicine well into the eighteenth century. Born in AD 129, he grew up as one of the gilded youths of the ancient city of Pergamon. The city was becoming hugely wealthy as a trading place as the Asiatic terminus of an important branch of the so-called Silk Road. Facing across the Mediterranean to Greece, yet being linked by its trading routes to India and China, it was an entrepôt for medicinal materials and a melting pot for medical theories. It was a natural place for doctors to congregate, and it developed a famous shrine to the god of medicine, Aesculapius. Attached to the temple of the god was a huge and immensely important library. There was also a famous school of gladiators, whose wounds provided, for those who could look at them, an immediate insight into anatomy. Galen's father was a prosperous architect, but dreamt that his son should be a doctor. That son was soon studying in Smyrna (modern Izmir), and in the Egyptian city of Alexandria. He returned to Pergamon in AD 157, and was appointed physician to the gladiators. He began to learn about blood, and the contents of the body. He learnt about healing too, and became such a success that five years later he moved to Rome. He seems to have been a flamboyant man, ambitious, funny and intelligent. He soon had a wealthy and influential clientele, and within seven years was employed by the ruling family.

Galen dissected goats, pigs and monkeys. He discovered, sometimes via public shows of his skills, how different muscles are controlled at different levels of the spinal cord. He showed that the brain controls the voice – severing one particular nerve could stop a pig squealing. He showed that arteries carry blood, disproving a widespread and ancient belief that arteries carried air. With no knowledge of how organs worked, or what caused diseases, he adapted the Eastern idea that illness was due to an imbalance of various body fluids; to Galen, these were the traditional black or yellow biles, phlegm and so on. However, he placed great importance on blood, thinking that most malfunctioning was simply an excess of it. Bleed the patient, as the Countess of Chinchon was bled, and the patient would be well.

Much anatomical work was also done in the Egyptian city of Alexandria, founded by Alexander the Great. A great trading city for medicinal drugs, it had an astonishing library that contained copies of all the great medical works of the past. The medical school was

OPPOSITE ABOVE Hippocrates was one of the most famous doctors of the ancient world. He is shown wise and serene in this fifteenth-century painting by Justus van Ghent.

OPPOSITE BELOW Ancient Pergamon, now Bergama in western Turkey, was birthplace to another famous doctor, Galen, and home to the most famous medical library in the Graeco-Roman world. The scrolls and books were later transferred to Alexandria, where they were burned. Here, Pergamon's ruins are shown in an epic but reasonably precise 1888 reconstruction by R. Bohn.

allowed the privilege of dissecting living humans, and much early work on the functioning of the nervous system was done there.

In the West, the fall of Rome and the isolationism of Christianity cut Western doctors off from both the classical past and contemporary developments in what was by then Arabic medicine. Anatomy ceased until the sixteenth century. As a knowledge of the human body gradually developed thereafter, following Vesalius's great work *De Humanis Corporis Fabrica* of 1543, some of the plants long used in the therapy of the whole human began to find new uses as they became associated with newly discovered organs, or the newly discovered functions of those organs. Plants became used specifically for lungs, heart, kidneys, liver, bladder and so on, and even moved from function to function, or organ to organ, as knowledge of the body, and of plants' effects on it, if any, changed. Some are still on the move. Some were especially mobile.

One such was liquorice, found in every single ancient medical source and used throughout the then known world for coughs, colds and asthma. Today, after many changes, it is scarcely more than a sweetmeat and a flavouring. Originating in the eastern Mediterranean region, liquorice (spelt licorice in the US) is mostly derived from the roots of *Glycyrrhiza glabra*, a shrubby member of the Leguminosae or pea family. It is listed in Babylonian cuneiform tablets of around 2600 BC, along with oils of cedar and cypress, myrrh, poppy juice, honey and nut galls. It is mentioned in the Ebers papyrus. Ancient Chinese and Ayurvedic sources also describe it. Around the Mediterranean, the Greeks copied its use from their wild and dangerous Scythian neighbours, nomadic horsemen who gathered its roots from around Lake Maeotis (the Sea of Azov), for Theophrastus, in the third century BC, had noticed that the Scyths used it for a range of chest complaints. Dioscorides, who named the plant glycyrrhiza (meaning sweet root), used it for the same range of ailments. Later, the Romans did too, while preferring to call the plant *Liquiritia officinalis*. The name later became corrupted as lycorys in thirteenth-century English, *lacrisse* or *lakriz* in German and Welsh. In France, it became *reglisse*.

Chewing the dried root releases its sweet flavour and many of the root's active substances, whilst filling the mouth with unpleasant fibre. The ancient Greeks discovered that boiling the root for an extended period, then evaporating the resultant sludge, yielded a sweet black fibreless rubbery extract. They traded it. The Romans traded it. It was traded throughout medieval Europe. It was well known in Germany in the twelfth century. In 1264, it was charged for in the Wardrobe Accounts of the English king, Henry IV. Druggists in the Frankfurt of 1450 bought it in from Italian apothecaries.

Living plants travelled too. By the mid-sixteenth century, there were fields of liquorice growing in Bavaria. It was an important crop in Britain, certainly at least since Turner's

The roots of liquorice (*Glycyrrhiza glabra*) were chewed for sugar and medicament, though liquorice was more often traded as a black and delicious extract. The plant is in fact an herbaceous perennial and not the charming small tree shown in this illustration from *Tacuinum Sanitatis* (c. 1385).

Liquiritia.

Nature. F. 7. h. in. 2°. melior exea non nimis grossa et coro
sa. Juuamenti. acens urtutem asperitati pectoris. nocumenti.
si nascitur interra creosa. remono nod. ueni plantata interra sab
losa.

Herbal of 1562. Gerard grew plenty in his London garden, and Nicholas Culpeper wrote: 'It is planted in fields and gardens, in divers places of this land and thereof good profit is made.' Pontefract in Yorkshire seems to have been the main producing area, though there were lesser centres at Mitcham and Godalming in Surrey, and at Worksop in Nottinghamshire. For reasons unclear, Pontefract's was the best. Growers there were so proud of their product that they stamped the small round lozenges made from the liquorice, then and now called Pontefract cakes, with the city arms from the late seventeenth century. Rival Dutch producers were also stamping their output, which commonly bore the designs of contemporary coinage, or were cast to look like scissors, keys or other fanciful shapes. In Europe, the range of liquorice's cures was already expanding fast. Culpeper used it for 'dry cough, hoarseness, wheezing and shortness of breath and all complaints of the breast and lung . . . consumption . . . the strangury [usually a result of an infection of the bladder] and heat of urine [probably gonorrhoea]'.

Although there is one American species of liquorice, *Glycyrrhiza lepidota*, it doesn't seem to have interested native Americans or colonizers. The Europeans couldn't do without their own sort, and some of the first settlers took it to North America. The enthusiastic seventeenth-century traveller John Josselyn gives the recipe for a beer strongly flavoured with elecampane, liquorice, aniseed, sassafras and fennel which he used to brew for the Indians when they had bad colds. Though he had trained as a physician in England, once in America, apart from administering his liquorice beer, Josselyn did no doctoring, spending his time travelling and marvelling at the New World. It may even be that the liquorice he found growing in the New England of the time was not even intended for its human inhabitants: Gervase Markham, a noted seventeenth-century authority on husbandry, gardening and farriery, wrote that it was an excellent horse cure.

With the increasing knowledge of anatomy, and the lessening of the taboo on dissecting dead humans, liquorice's effect on other organs than the lungs became apparent. The plant contains chemicals very close in action to the hormone cortisone, which calms inflammations. Herbalists had noticed this by the late sixteenth century; Gerard says that he used it to heal ulcerated kidneys and bladder. Later, a Dutch pharmacist, observing the traditional pulmonary use of liquorice juice by the people of southern Italy, noticed that it also helped cure gastric ulcers. Liquorice's active substance, glycyrrhizin, is very similar both chemically and pharmaceutically to carbenoxolone, a chemical now used for mouthwashes, but which was until recently an important treatment for gastric ulcers.

In spite of the changes in liquorice's use among advanced doctors, the ancient tradition died very slowly among the people. In cold damp northern Europe, where colds, catarrh and other pulmonary problems were rife, liquorice was considered an essential commodity. It was though, for the poor, expensive. Soon, any ordinary hedgerow plant that had long and sweetish roots and a roughly similar flavour was called 'wild liquorice'. The restharrow (*Ononis* spp.), some of the vetches (*Astragalus* spp.), even sweet cicely (*Myrrhis odorata*) were all given this appellation, and used as substitutes, even adulterants, of true liquorice. Only sweet cicely has chemical components that are at all similar. In the great industrial cities of the nineteenth century, with their terrible 'smokes' and consequent lung diseases, liquorice became more important still. One contemporary homemade medicine for bronchitis sounds good:

Take a large teaspoonful of Linseed, 1 ounce of Liquorice root, and ¼ lb. of best raisins. Put them into 2 quarts of soft water and simmer down to 1 quart. Then add to it ¼ lb. of brown sugar candy and a tablespoonful of white wine vinegar or lemon juice. Drink ½ pint when going to bed and take a little whenever the cough is troublesome. (N.B. – It is best to add the vinegar to that quantity which is required for immediate use.)

Liquorice was also much in demand for disguising the taste of nasty medicines, especially acrid or bitter ones such as mezereon, quinine or cascara. It is still used in comfits such as 'meloids', where it somewhat disguises the taste of the menthol and capsicum that reduce the symptoms of the common cold by relaxing the nasal tracts.

It wasn't only in Europe that liquorice was important. Dried roots, or the black 'paste', had moved eastwards along the Silk Road. In China, mixed with the poisonous peony, it was used for abdominal pains. But its uses increased mightily, until it became eventually second only in importance to the fabulous ginseng. Liquorice – twigs, leaves and flowers – was, like that ultimate 'shadow plant', believed to be a rejuvenator if consumed over a long period of time. It was believed to protect the taker against poisons, including poisoning by aconite and poisoning due to the overuse of ephedrine. (The juices from *Ephedra* species were a widely used sex stimulant.)

Liquorice became important in Arabic medicine too. The famous medical encyclopaedia compiled by Ali ibn Rabban al-Tabari (838–870 CE) describes how that famous physician used it to strengthen the stomach and to ensure the good health of his patients. He prescribed 'black myrobalan powdered in butter, mixed with dissolved plant sugar extracted from the licorice'. In India, Ayurvedic medicine eventually suggested that it was excellent for the voice (presumably clearing the throat) or mixed into warm mild milk as a heart tonic. Following Unani, or Western, medicine, Indian physicians recommended it for bronchitis, colds, cough and debility in general. They also used it as an emetic in large doses, as a cure for hyperacidity, inflammation and laryngitis, sore throats, ulcers and 'urination pain', as a laxative, and as an aid to mental calming. The list of its Ayurvedic uses goes on: blood purification, abdominal pain, nourishment for the brain (it was thought to increase cranial and cerebrospinal fluid) and a tonic for complexion, even hair and eyes.

For something that we eat so casually when buying a box of liquorice allsorts or a pack of liquorice Catherine wheels, liquorice has an astonishingly varied and ancient history. When eating it we are also partaking of a quite astonishing list of chemicals: liquorice contains a huge repertory of active substances. Delicious though all this is, in high doses of up 12g or more it can cause hypertension, hypokalaemia, and sodium and water retention. It also contains a chemical that mimics the female sex hormone oestrogen. Perhaps that is why, according to one ancient belief, a woman who chews a piece of liquorice root feels sexy. It

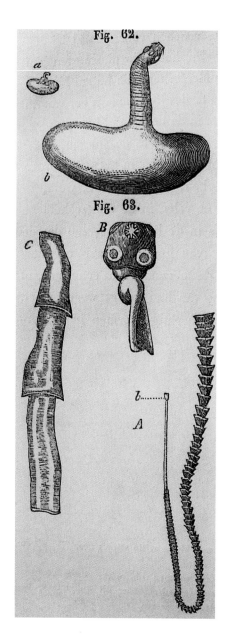

Fig. 62.

Fig. 63.

Wood carvings by W.H. Stam (from Jan van der Hoeven's *Natural History of the Animal Kingdom*, 1857) illustrate various stages in the life of a tapeworm, whose barbed head anchors itself high up in the gut so that it can absorb nutrients over its entire length.

can also mimic hormones that regulate appetite. Manual labourers once chewed pieces of liquorice root to make them feel less hungry.

That liquorice has moved, during its 3,000-year history, from the lungs to the belly, is not surprising, for the alimentary canal therein is easily affected by plants. Over the millennia, we have found huge numbers of them that can variously loosen it up, close it down, stop it bleeding, dispel its gases and discharge the various toxic materials that were once held to accumulate within it and cause distempers. Until recent times in the Western world, some of these alarming maladies of the innards were caused not by unbalanced humours, or difficulties with yin and yang, but by worms. The roundworms and tapeworms we now occasionally see in our pets' vomit or faeces once also inhabited us, in some quantity. There were, and are, many types. Some roundworms, or ascarids, can grow to over a foot in length. Strong and mobile, they often ascend through the stomach to the throat. In the days when people were affected by them, if the human host was deep enough asleep not to retch, they could be found coiled and cold amongst the bedclothes in the morning. Infestations of the smaller sorts of roundworms, particularly whipworms, could, and can, be very debilitating. Whipworms burrow into the gut lining to absorb nourishment directly from the blood. In large numbers, they make the host's intestines bleed copiously, colouring the faeces. Ascarids and roundworms are mostly picked up from faecal contamination of food, drink or fingers. Of the types of tapeworm that can inhabit us, some can grow to immense lengths: worms thirty feet long are commonplace. One human can entertain several. Tapeworms usually need an intermediate host to complete their life cycle; the sorts that can infest us commonly need pigs, chickens or various game for one stage of

Many species of *Artemisia*, including this one, *A. absinthium* – from Sowerby's *English Botany* – were used to expel worms from the innards, often taken infused in wine – a mix that gave us delicious and sometimes dangerous drinks such as absinthe and various vermouths.

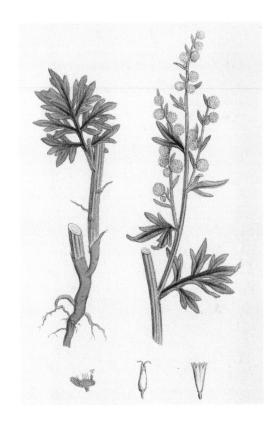

their life cycle. They reach us if the host animals are not sufficiently cooked. All were once common.

At least by the sixteenth century, it had become clear that the wormwoods consist of a number of species. All belong in the genus *Artemisia*, pungent-leaved shrubs or herbaceous perennials, and an annual or two. Many are still in the garden: the silver-leaved *Artemisia pontica* (called old warrior); the shrubby and grey-green *A. abrotanum* (southernwood, sometimes lad's love on account of its supposed aphrodisiac powers); *A. lactiflora*; *A. absinthium*; and of course the familiar *A. dracunculus* (dragons or tarragon). As the need for vermifuges lessened, many found dramatic new uses for themselves, often moving from the gut – in particular the stomach – on to other organs. *Culpeper's English Physician and Complete Herbal* says of artemisia: 'to expel worms ... [the leaves] made into a light infusion, strengthen digestion, correct acidities, and supply the place of gall, where, as in many constitutions, that is deficient'. Gerard says pretty much the same: that it 'voideth away the wormes of the gut ... strengthen and comforteth the stomacke ... yeeldeth strength to the liver also cureth the yellow jaundice'. Worms couldn't cope with the sesquiterpenes that give the leaves their sour, pungent smell. Also, the substances that give artemisias their odour have a stimulating effect on some of the specialized cells of the lower part of the stomach called the antrum, increasing their secretion of gastric juices. Artemisias thereby improve digestion. They also stimulate the gall bladder to squeeze its bile into the gut, aiding later stages of digestion too.

Following the bon vivants of Propontis and Thracia, *Artemisia absinthium* and others became used in other parts of Europe, combined with wine, as a digestive. The combination was often taken before a meal, or even throughout the day. In 1772 Sir John Hill (*c.* 1714–75) recommended the closely related common wormwood (*A. vulgaris*) in many forms. He writes in *Eden; or, a Complete Body of Gardening* of 1757:

The Leaves have been commonly used, but the flowery tops are the right part ... One ounce of the Flowers and Buds should be put into an earthen vessel, and a pint and a half of boiling water poured on them, and thus to stand all night. In the morning the clear liquor with two spoonfuls of wine should be taken at three draughts, an hour and a half distance from one another. Whoever will do this regularly for a week, will have no sickness after meals, will feel none of that fulness so frequent from indigestion, and wind will be no more troublesome; if afterwards, he will take but a fourth part of this each day, the benefit will be lasting.

This recipe is a water extraction, leaving in the plant all the essential oils, which need alcohol to dissolve them. Hill also waxed lyrical about the use of wormwood in fortified wines, which had plenty of alcohol for this purpose: 'Wormwood wine, so famous with the Germans, is made with Roman Wormwood … it is a strong and an excellent wine, not unpleasant, yet of such efficacy to give an appetite that the Germans drink a glass with every other mouthful, and that way eat for hours together, without sickness or indigestion.'

These German infusions, called 'vermuts' or vermouths, developed fast. At the end of the century, a new patent medicine appeared in Switzerland. Its origin is often attributed to a man called Dr Pierre Ordinaire, and to the year 1792. Used as a cure-all, it was speedily nicknamed 'La Fée Verte'. It combined most of the *Artemisia* species during some stage of its preparation. It also included melissa or sweet balm (*Melissa officinalis*), angelica (*Angelica archangelica*), hyssop (*Hyssopus officinalis*) and large quantities of anise (*Pimpinella anisum*) and fennel (*Foeniculum vulgare*). It ended up being intensely flavoured, immensely alcoholic and, in the bottle, a glittering light peridot green. Diluted with water in the glass, it turned opalescent as the dissolved oils came out of solution. The medicine slowly caught on. It was called absinthe.

The recipe was brought to France in the early 1800s by a Major Dubied, who purchased it from the sisters Henriod. It is uncertain how it became the sisters' property. Dubied's son-in-law, Henri-Louis Pernod, set up a factory for its production at Pontarlier, France, in 1805. The major seems to have ensured that the cure-all was sold to the army. It was used by French troops fighting in Algeria from 1844 to 1847, who needed something to protect them against fever and fatigue. Many soldiers developed a taste for it, and wanted to keep it near to hand when they returned to France. Word had spread that it was an aphrodisiac, perhaps because it contained southernwood (*Artemisia abrotanum*). There were also whispers that it was a mild hallucinogen. There were vaguer whispers still that it was addictive. In other words, it was dangerous. If so, avant-garde artists and their hangers-on, the racier parts of high society and their hangers-on too, all wanted to try it. The absinthe market grew fast. It was expensive at first, so other distilleries were set up to make cheaper versions. There were substitutions in the plants used and more dangerous ones crept in. In the mid-1870s, phylloxera attacked the vineyards and destroyed the vines, and wine and brandy prices rose drastically. Absinthe manufacturers turned from using brandy to grain alcohol, which was

A splendid poster of *c.* 1900 advertising Absinthe Blanqui proclaims the restorative and stimulating powers of absinthe. No wonder sales rocketed.

abundant and still extremely cheap. Suddenly, absinthe was available to everyone. Competing brands fuelled public interest with advertisements suggesting that it was merely a healthy herbal tonic. Others assured the drinker the wildest of nights. No cheap brand admitted that its product was coloured no longer with fresh herbs but with poisonous copper salts. Some of the posters were exceptionally elegant. Oscar Wilde, Ernest Dowson, Pablo Picasso, Arthur Rimbaud, Paul Verlaine, Aleister Crowley and Charles Baudelaire all sat in cafés and drank it. Van Gogh may have sliced off his ear partially under its influence. Toulouse-Lautrec made a special concoction called *atremblement de terre*, or earthquake, which combined absinthe and cognac. Some lesser folk combined absinthe with red or white wine instead of water. Purists drank it neat.

In 1874, the French consumed 700,000 litres of absinthe. By 1910, they were drinking 36,000,000 litres. It was becoming a social threat. It became associated with epileptic seizures, orgiastic behaviour and sexual diseases, corrupted artists and criminals. Sensational murders were supposedly committed under its influence. It was soon banned in Holland, Belgium and Brazil. It had also taken hold in America. Horrified by both alcohol and licence, United States health officials imposed a ban on the drink in 1912, though it continued to be available if the devotee knew what sort of hair tonic to buy. France finally banned it too in 1915. However, French aperitif makers, including Pernod, were creative. They designed new aperitifs that did not include any of the supposedly dangerous wormwood, but had far more anise and just as much alcohol. The chemical usually blamed for absinthe's exotic reputation is alpha-thujone. It is widespread in the plant world, and is in particular found in herbs such as sage (*Salvia officinalis* – though, as we shall see, there are some very strange salvias) and tansy (*Tanacetum vulgare*). Fennel and anise have plenty too. The new vermouths and types of pastis contained almost as much alpha-thujone as the strongest absinthes. Alpha-thujone, administered in large enough doses, can indeed cause epileptic fits, at least in laboratory rats. The rats in the experiments also exhibited a host of other odd symptoms. The dosages, though, were equivalent to a human drinker having thousands of glasses of absinthe. Perhaps the sexual arousal, the hallucinations and the deep mysteries of absinthe were more in the drinkers' minds than in the glass, and yet another aspect of the placebo effect.

It seems quite possible that the wormwoods will continue to affect our lives. One species, *Artemisia annua*, has been rediscovered as a medicinal plant in China. Its use in malaria has already been mentioned. Recent researches show that artemisia killed, *in vitro*, virtually all human breast cancer cells exposed to it, within sixteen hours. The plant may yet have more to offer.

Of our bodies' organs, neither stomach, nor liver, nor alimentary canal, all vital to our lives, has ever been thought of as expressing anything much about our deeper selves. In contrast, as the heart was, and popularly still is, seen as the seat of the emotions, so are the eyes regarded as windows to the soul. Many plants have been associated with their maintenance and healing. The strangest story of all is to do with a plant that was believed to make them more beautiful. The botanist, and doctor to Emperor Ferdinand I, Pietro Andrea Matthioli, usually known as Matthiolus (1501–77), wrote that fashionable

Pietro Andrea Matthioli was a physician from Siena. His great work was a commentary on the books of Dioscorides which, published in 1544, became the most widely read medico-botanical book of the period. He died of the plague in 1577.

Venetian women used a water infusion of the plant to make their eyes look darker and more lustrous. The name for the plant was therefore 'Bella-donna', or beautiful lady. It worked by dilating the pupils of their eyes, making them look darker and deeper, and also duplicating one of the signals of sexual arousal. But they, and the Spanish *majas* who were also reputed to use the plant, were taking a dangerous path, one not unknown in the long history of cosmetics. A much older name for the plant was Atropos, after the one of the three Fates who wielded the scissors of death. The Latin name for the plant is *Atropa belladonna*.

Atropa belladonna is native to a swathe of territory that runs from Britain, across central and southern Europe, halting only at its eastern limits in Iran. Such has been its importance in human affairs that it now has an almost worldwide distribution. It is rather grand to look at, with a big rosette of grey-green leaves and long branching and leafy flower spikes. The flowers are usually a strange violet-green, and the subsequent fruits look like glistening black cherries.

Even for those not interested in beauty, who merely wanted to see the world in focus, the plant remained in use until recent times. The ability of the plant to dilate the pupil is important in ophthalmology. Until the last few decades, opticians used it regularly as eye drops before carrying out an eye test. It is still used during eye surgery. But its common names give away its darker side: deadly nightshade, or dwale, after an ancient German word for delirium. One of the plant's constituents is atropine, which is poisonous. All parts of the plant contain it, but the glistening black fruits are especially poisonous. It takes only a few of them to kill.

Harvey Wickes Felter, a doctor from Cincinnati, Ohio, wrote in his *The Eclectic Materia Medica, Pharmacology and Therapeutics* of 1922:

Small doses occasion dryness and constriction of the throat, with possibly disordered vision and such unpleasant head symptoms as vertigo and confusion of ideas ... Large and toxic doses greatly augment the dryness and dysphagia and giddiness, the patient reels or staggers when he walks, there is great thirst, and sometimes drowsiness and nausea and vomiting occur ... Vision is either lost, or indistinct and double. The rate of the pulse may be doubled and the volume is full and hard ... The eyes are brilliant and staring ... A peculiar active delirium accompanies and is of an illusional and loquacious character. The victim ... sees visions, entertains spectres, has fancies and hallucinations, and other

Atropa belladonna, poison, cosmetic,
hallucinogen of witches, painkiller and much
more: this 1796 illustration of the sinister
plant, with its seductively glistening berries,
is by Friedrich Justin Bertuch.

phantasmagoria, and gives way to laughter and gayety; again the cerebral disturbance may be of a wild maniacal type, with furious delirium and fighting propensities. Loss of speech often occurs early, though repeated movements of the tongue and lips indicate the efforts to articulate ... Finally, with (rarely) or without convulsions, occurs a complete abolition of function, stupor sets in ... and paralysis closes the scene in death, which results chiefly from respiratory paralysis. Should recovery take place, the patient seldom recollects any of the circumstances of his illness.

That such a strange and dangerous plant should ever have been the remarkably successful medicine it became takes some explanation. In ancient Greece, it seems to have been used, perhaps combined with alcohol and the juice from ivy leaves, during Bacchanalian orgies in which the Bacchantes – female devotees of the god Bacchus – danced and tranced. Its effects made them randy, violent and dangerous, and an enticing threat to the male onlookers. Theophrastus thought the plant a form of mandrake, and it is indeed closely related. It was supposed to have been the plant that poisoned the troops of Marcus Antonius during the Parthian wars. During the Dark and Middle Ages it disappeared into hospitals, or into the hidden world of pagan ritual and shamanism. It reappears in the mid-fifteenth century as 'Solanum minus', vanishes, then appears again in the next century. Authors all urge caution.

In 1578, Henry Lyte (*c.* 1529–1607), in the *Nievve Herball*, his translation of Rembert Dodoens' *Crüÿdeboeck*, tells growers 'to be carefull to see to it and to close it in, that no body enter into the place where it groweth, that will be enticed with the beautie of the fruite to eate thereof'. Gerard calls the plant 'Solanum lethale', and 'Sleepy Nightshade'. He warns the reader, after recounting three cases of poisoning from eating the berries, to 'banish therefore these pernicious plants out of your gardens and all places neare to your houses where children or women with child do resort, which do oftentimes long and lust after things most vile and filthie; and much more after a berry of bright shining black colour, and of such great beautie'. Many alchemists were interested in it, and even early scientists such as Gianbattista della Porta got into trouble with the Inquisition for describing experiments with it. During the sixteenth century and the next, it became more and more openly associated with witches. According to stories of the time, the plant belongs to the devil, who goes about trimming and tending it in his leisure. The only day he forgets to do so is Walpurgisnacht, when preparing for the witches' sabbath.

Witches, and wizards, were said to rub a mix of *Atropa belladonna*, hemlock, mandrake and henbane, pounded with bear's grease, known as 'sorcerers' pomade', on to their thighs and genitals before setting off on their broomsticks. Even herbalists recognized that belladonna's alkaloids can be absorbed through the skin. Gerard suggested applying leaves to the brow to reduce a violent headache. The application of too many leaves would have resulted in intoxication and hallucinations. It really does seem to have dulled pain – hence its later and liberal use in nineteenth-century American sticking plasters. Though the plant was vilified in the seventeenth century, the Age of Reason took to it enthusiastically. Its effects were the subject of treatises by doctors all over Western Europe. Finally, in 1819, its active chemical, atropine, was discovered by Brandes. Pure atropine was available by 1833. Across

The absorbent pads of modern sticking plasters may contain mild bactericides, but these 'Belladonna plasters' were supposed to contain enough atropine to dull local pain. Some brands were found, when analysed, to contain none. Some had dangerous amounts.

the Atlantic, enthusiasm boomed. It first appeared in public in *The Pharmacopeia of the Massachusetts Medical Society* in 1808, then in the 1821 edition of *The American New Dispensatory*, and six years later in the *Eclectic Dispensatory*, published in Philadelphia. Atropa, or its alkaloids, found its way into suppositories, enemas, plasters and injections, all recommended by enthusiastic doctors. Even its name was useful in selling medical supplies. The *United States Dispensatory* drily observes that 'machine-spread "Belladonna plaster" can be found on the market, which was admitted by a representative of the manufacturer, to contain no Belladonna whatever'. Of the many salves, ointments, rubs, liniments and so on that really did contain it, Dr Harvey Wickes Felter wrote in his *Eclectic Materia Medica* of 1922: 'Ointment of belladonna and the related liniment are extremely useful in local inflammations and swellings, having a wide range of efficiency. Thus they may be applied to painful and swollen joints, forming abscesses, incipient and recurrent boils, buboes, haemorrhoids and fissures, inflamed glands, and in neuralgia, chronic rheumatism, lumbago, myalgia, pleurodynia, the chest pains of pulmonary tuberculosis, and in acute mastitis.' However, he added the following caution: 'A small amount of an ointment containing only three grains of atropine produced death in two hours.'

Amazingly, belladonna was even combined with other poisonous plants. Felter liked it combined with aconite for a 'severe sore throat with redness, rawness, swelling, intense soreness, difficult swallowing, and dryness of the throat, with or without fever'. The same dread mix was also used for 'tonsillitis, especially of the quinsy type, and in pharyngitis and faucitis'. More surprisingly still, it was, in both Europe and America, much given to children. Felter again: 'Chicken-pox does not so often demand it ... of small-pox it is claimed to be a most certain aid ... We rely upon it absolutely in scarlatina, and the more malignant the type the more it is indicated. We do not recall a case of scarlet fever in which we have not employed it ... It serves much the same purpose in measles, and helps also to control the cough ...' As we shall see, exceptionally poisonous plants seemed to exercise a fascination on many nineteenth-century doctors. Many thought them a perfect cure, in small doses, for practically everything. It is difficult to know where the root of the fascination lies. It is as if in some doctors there is a strong 'shadow side' to the urge to heal: Janus humans to match Janus plants.

When early-nineteenth-century chemists examined belladonna, they discovered several powerful alkaloids. One they called atropine after the plant's Latin name of *Atropa belladonna*. The others they called hyoscyamine and scopolamine. The first of these is named after *Hyoscyamus niger*, the henbane. The second is named after a beautiful Japanese woodland plant now in the genus *Scopolia*. *S. japonica* has horizontal knobbly runners, which each spring send up brilliantly green leafy stems from which hang tubular purple-brown and cream bells. The roots are often used as a substitute for belladonna, though they contain less atropine. Scopolamine and its associated plant have only been much used in the West since the late nineteenth century. In the early twentieth century scopolamine was frequently combined with morphine to induce a trance called 'twilight sleep' that was supposed to lessen the pain and the mortality of childbirth. More recently it has been used as a 'truth drug', and even for the unpleasant business called brainwashing. Some travel sickness pills also contain it. The plant commemorates poor Johann Anton Scopoli, an Austrian doctor captured by Italian pirates and forced to work for them as their surgeon. He got hold of the plant through Venetian apothecaries and in his *Flora Carniolica* of 1766 puts it in the genus *Atropa*. Dying in poverty in 1788, he was never to learn of his fame.

As work on the anatomy and workings of the eye continued, at first, only European plants like belladonna were associated with healing it. In the nineteenth century, when Westerners had access to the flora of the entire globe, more exotic plants began to be used. The most famous of these came from the rainforests of South America: two species of *Pilocarpus*, *P. jaborandi* and *P. microphyllus*, which contain the alkaloid pilocarpine and from the dried leaves of which the drug jaborandi is made. Both plants are large-leaved understorey shrubs from the Amazonian jungles of Brazil and Paraguay, where they were widely used by the indigenous tribespeople. In 1570, Gabriel Soares de Souza reported noticing that the Guarani Indians used jaborandi to treat mouth ulcers. In the 1630s two Dutch West India Company doctors saw other Indians using it for colds and flu, and as a remedy for both the newly imported gonorrhoea and various poisons. When using it for the common cold, the Indians wrapped the sufferer in blankets to raise his temperature. Administering an infusion of jaborandi rapidly produces huge amounts of sweat, saliva and urine. The Indians believed that the heat and sweat expelled the disease. Unlike quinine, it didn't cause much interest in Europe. Then, in 1873, a young Brazilian, Symphronio Coutinho, came to Paris to study medicine, bringing some of the leaves with him. The first scientific studies began to appear the following year.

Three years later, jaborandi had become what every new and exotic drug has become since the dawn of medicine: the love object of the adventurous doctor, used as a cure for everything. In jaborandi's case it was tried as a cure for 'fever, stomatitis, enterocolitis, laryngitis and bronchitis, bronchiectasis, influenza, pneumonia, hydropericarditis, hydropsy, psoriasis, intoxications, neurosis and renal disease'. It was even used in malaria, one report running:

... in a strong man, the bowels had not been moved for three days, nor the urine voided for eighteen hours. The temperature was 107 degrees, and the pulse 140, full and bounding. ... One-half grain of pilocarpine hypodermically, caused salivation in three minutes, perspiration stood on the neck and forehead in great drops, the face and skin became extremely red at first and pale as the perspiration

advanced. He immediately passed a large quantity of dark-colored, highly offensive urine. He vomited, and had a movement from the bowels, large and copious. The doctor claimed that more was accomplished by this one dose of medicine in one hour in the way of elimination, than he could have accomplished otherwise, in forty-eight hours.

It was also used to ease painful gall bladders and difficult births, and even as a hair-restorer. Oddly, the new hair was supposed to be in its original colour, not grey. It is still sometimes found in modern shampoos.

However, a year or two after all that, jaborandi found its true organ: the eyes. It was discovered that pilocarpine lowers the fluid pressure inside the eyeball by acting directly on cholinergic receptor sites; this is useful in cases of glaucoma, in which high pressure is a characteristic. Pilocarpine became, and remains, the standard drug used for initial and maintenance therapy in certain types of primary glaucoma. But jaborandi is still developing. When it causes sweating, it also opens the skin pores and increases blood flow to the tiny local blood vessels. This enables other drugs to penetrate the skin and act locally rather than generally, as they would if taken orally or injected. The wild bushes of the *Pilocarpus* species

have become valuable. The increasingly few natural populations are now on the official list of endangered plants in Brazil, and seed collections of jaborandi have had to be established at various Brazilian institutions to protect its future. However, though nowhere near as poisonous as belladonna, jaborandi also has two faces. Large doses can reduce the pulse rate to dangerously slow levels. But before death occurs by that route, the lungs are usually flooded with a huge overproduction of mucus and the patient drowns instead.

The jaborandi grows in wildest Amazonia. The belladonna plant is hardly

Pilocarpus jaborandi, 1895, by Matilda Smith, from Curtis's *Botanical Magazine* (1896). Growing in the rainforests of the Amazon, it is used by native Indians to treat mouth ulcers, colds, fevers and kidney stones, and as a antidote to poisons. In the West, clinical research into its properties continues, especially as when applied to the skin it allows other topical pharmaceuticals to enter the bloodstream.

seen wild in European or American hedgerows. Neither is much of a threat to the general public. Yet some common native hedge and woodland plants, admired and cherished, are Janus plants too, every bit as dangerous but useful medicinally.

The common foxglove (*Digitalis purpurea*) is widespread in British hedgerows and woodlands; there are various other species in the rest of Europe and Asia. Each holds a slightly different range of chemicals. Two have been used medicinally. *D. purpurea* contains digitalin, digitoxin and a range of glycosides, as well as other ingredients. The grey-green-leaved and yellow-flowered *D. lanata* contains digitoxin too, and other compounds not in *D. purpurea*. With such a bewildering array of compounds, each varying throughout the plant's season of growth, it is no wonder that confusion reigned for so long about what effects foxglove actually has on the body. One thing was, though, widely known: foxglove is dangerous.

In the Middle Ages foxglove was used as a cure-all, or a 'herb of grace'. However, when the strange 'doctrine of signatures' was applied to it, it became clear that its use should change. This doctrine proposed that a plant's appearance enabled the herbalist to decide what it might cure. In the common foxglove, the flowers are splendidly spotted and speckled along the outer margins of the corolla, and it was believed that the spotted flowers made it appropriate as a treatment for diseased and spotted lungs. Many plants similarly marked, whether on flowers or on leaves, like that other 'lung' plant *Pulmonaria officinalis*, were given a similar use. The doctrine gave rise to some curious couplings. William Coles, in his *Nature's Paradise* (1650), thought that walnuts were good for curing head ailments because the nuts look a bit like the brain. Another of his examples was: 'The little holes whereof the leaves of Saint Johns wort [*Hypericum perforatum*] are full, doe resemble all the pores of the skin and therefore it is profitable for all hurts and wounds that can happen thereunto.' The idea may well be ancient, for aspects of it appear in both Chinese and Ayurvedic medicine. In Europe, it was loudly promulgated by Paracelsus, a Swiss doctor whom we shall meet in the next chapter, who thought that God gave everything in nature a unique healing power and the divine clue given was in the appearance of each plant or substance. Paracelsus wrote in Latin, but his ideas reached a wider audience in the early 1600s. An important follower was Jakob Böhme (1575–1624), a master shoemaker in the small town of Görlitz, in Germany. He had a profound mystical vision in 1600, in which he saw the relationship between God and man, and man and plants. He began describing these mystical inter-relationships in his native language, not Latin, and his pamphlets were widely read and translated, thus furthering Paracelsus's views all over Europe.

Parkinson, in his *Theatrum Botanicum* of 1640, was not convinced about foxglove's use for lung complaints, noting it only as a wound herb. Nicholas Culpeper was of much the same view. He writes: 'The herb is frequently and familiarly used by the Italians to heal any fresh or green wound, the leaves being but bruised and bound thereon; and the juice thereof is also used in old sores to cleanse, dry and heal them. An ointment of it is one of the best remedies for scabby head that is.' However, early in the next century, the old uses were combined with the new ones suggested by the doctrine of signatures. Even by 1710, one herbalist wrote that it was 'an extraordinary good wound-herb, prevalent against the King's

Evil [or scrofula, a tuberculosis of the skin], and may be used instead of gentian ... It is a specific which transcends all other vegetable medicaments for the cure of consumptions; cleaning and healing after an admirable manner ulcers of the lungs ... when all other medicines have failed, and the sick been esteemed past cure ... I have known it do wonders.' Some of the dosages suggested were alarming. Foxglove has what is called a low therapeutic ratio. That is, the difference in dosage between saving and slaying is small. The rates at which it is absorbed and excreted by the body are also immensely important.

In George Eliot's novel *Silas Marner*, there's something about the central character, a strange and alienated man, that makes those around him wary. 'And where did Master Marner get his knowledge of herbs from – and charms too, if he liked to give them away? ... Marner had cured Sally Oates, and made her sleep like a baby, when her heart had been beating enough to burst her body, for two months and more, while she had been under the doctor's care. He might cure more folks if he would; but he was worth speaking fair, if it was only to keep him from doing you a mischief.'

Marner used foxglove. 'One day, taking a pair of shoes to be mended he saw the cobbler's wife seated by the fire, suffering from the terrible symptoms of heart-disease and dropsy which he had witnessed as the precursors of his mother's death. He felt a rush of pity at the mingled sight and remembrance, and recalling the relief his mother had found from a simple preparation of the foxglove, he promised Sally Oates to bring her something that would ease her since the doctor did her no good ...' The novel is set around 1810, and the cure in question is remarkable. Marner's mother must have been well informed, for the common foxglove was only just switching uses yet again. She would certainly have known its traditional uses as a wound dressing and 'cure' for skin diseases. Yet, by the early nineteenth century, it was only just being used in very advanced medical circles as a cure for oedema, or dropsy, because of its effect on dropsy's underlying heart condition – a weak heart gives rise to oedema in the tissues, particularly in the legs.

Though Galen of Pergamon had shown that the heart was the pump of the body, it was a thousand and more years before the way the heart actually worked was discovered. A couple of centuries after that, its malfunctioning was shown to be linked to all sorts of other symptoms in the body. The novel was written in 1861 when the malfunctions and their symptoms were well understood, but Silas's awareness of them in 1810 is almost certainly an anachronism.

Foxglove was known to be a plant of the night as well as of the day, and everyone, apart from Silas Marner, knew that it was dangerous. There is, in fact, no such thing as a 'simple preparation' of *Digitalis purpurea*. The digitoxin in foxglove is absorbed rapidly, and slows a racing pulse – as it did Sally's. The body tissues at once begin to break the digitoxin down, ensuring that it is eventually eliminated. However, digitoxin has a half-life of about a week: it takes a week for half of one dose to be excreted, another week for half of the remainder to go and so on. If a patient takes even moderate doses at too frequent intervals, it accumulates in the tissues until toxic levels are reached. If dosing is done carefully, a steady state does not occur until after about five half-lives – that is, about five weeks.

William Withering (1741–99) was the first man to realize how *Digitalis purpurea* could be

The common foxglove (*Digitalis purpurea*), once used to treat tuberculosis and general wounds, found its real use in the mid-nineteenth century: its cardiac glucosides have a powerful effect on the heart. It and *D. lanata* are amongst the few medicinal plants grown as farmed crops for their contents. This illustration is from Curtis's *Botanical Magazine*.

given safely. The son of an apothecary, he studied medicine at the University of Edinburgh, and set up as a doctor in Birmingham. He became a member of the extraordinary Lunar Society of that city, founded in 1766 by Matthew Boulton and Erasmus Darwin (Charles's grandfather), along with their doctor, William Small. When Small died, Darwin wanted another doctor member, and invited Withering to join. The Lunar Society meetings were held once a month at full moon, not with witchery in mind, but so that the members might more easily find their way home after dinner. Conversation was as intellectually brilliant as at London's Royal Society, but, far away from that centre of scientific advance, members could discuss topics in an unconventional way that few fellows of the Royal Society would have risked. Withering was already well known. His practice was so successful that by the time he was forty-six he was renting the handsome Edgbaston Hall and had become one of the richest doctors outside London. In 1775, one of his patients was dying of dropsy.

Withering thought the case was hopeless. The patient, unwilling to give up, took a gypsy remedy, and got better. Withering, not at all put out, hunted down the gypsy and found that the cure's vital ingredient was foxglove. He was intrigued. He tried every bit of the plant, and after experimenting on 163 of his patients, found that dried powdered leaf, turned into a tea, was the best way of administering the drug. Around 1782, he realized that the convention of frequent dosing, which he had originally followed, led to high levels of foxglove in the body and patients became very ill indeed. He reduced the frequency of dosing to once or twice a day, and for only a few days at a time. The patients gradually got much better. Ten years later, in 1792, and after much discussion at the Lunar Society, he published a treatise on the treatment of dropsy (oedema, or accumulation of water in the body tissues) with foxglove. In so doing, he also described its effect upon the heart. It was a landmark in medicine. The plant had found its true organ.

George Eliot probably knew about Withering's work and the implications for the dosage of foxglove. Correctly, her creation Marner doesn't. He has only old-fashioned folk knowledge: 'He had inherited from his mother some acquaintance with medicinal herbs and their preparation – a little store of wisdom which she had imparted to him as a solemn bequest ...' His mother was probably dead long before Withering published the results of his experimental work on administering foxglove.

As turns out to be usual with such potent and dangerous herbs, some doctors experimented with the use of foxglove in ways which seemed to breach the Hippocratic oath. One such was Dr John Coakley Lettsom. He was born in the Virgin Islands, West Indies, in 1744. Sent to England when he was six, after a circuitous route he ended up as a medical graduate. He studied first under various London luminaries, then at Edinburgh University. A man of talent, he was hugely successful, and was one of the founders of the prestigious Medical Society of London. 'Lettsomian Lectures' are still given. A progressive, he warmly supported Edward Jenner (1749–1823) and his claims for vaccination against smallpox. He wrote a study of the mangel-wurzel and a history of the tea tree, for which latter he was complimented by Linnaeus. Yet, in spite of all that, he was at the very least cavalier with the well-being of some of his poorer patients. Writing after the appearance of Withering's researches, and foxglove's swift move from the lungs to the heart, he says:

In the exhibition of the Digitalis purpurea, the first effect I have observed, is rendering the pulse slower than in the natural state of the patient. Thus persons whose usual standard may be 60 to 70 ... have had the pulsations reduced to 56, or even less in a minute ... If the dose be increased till nausea or sickness is excited, the strength of the patient is still more weakened; the sickness resembles sea-sickness accompanied with a confused aching and heaviness of the head. The patient at this period remarks that he perceives flashes of light frequently pass across his eyes, and sometimes balls of Fire in the room. An increase of the dose after this produces vomiting, and sometimes purging also: he complains of increased headache, or rather of confusion and giddiness; instead of flashes of light almost all objects he views appear brilliant, and his friends who visit him seem to be surrounded with a blaze of fire; his memory is imperfect, and upon attempting to walk, he reels and staggers like a person Intoxicated. The dose that brings on these effects, gradually produces confused vision, and at length almost total blindness, which I have known to continue in some instances upwards of a month after the medicine has been omitted. During this time he complains in a particular manner of a throbbing pain in the balls of the eyes, and a sense of fulness and enlargement of them, as if the globes had become too big for the sockets, and were grown out of their natural size. In two cases that I heard of the limbs, particularly the lower extremities, were seized with tremors; and from some cause or other both these patients died suddenly, in a manner most resembling apoplexy.

Terrifying. Foxglove also found itself in use as an anaphrodisiac in women, prescribed by some nineteenth-century doctors. In men too, it was supposed to inhibit erotic excitement, whether this excess was due to 'excitable temperament, sedentary life, stimulant regimen, or the privation or excess of venereal pleasures etc'. No doubt patients of both sexes found it difficult to feel 'erotic excitement' if they had a pulse rate of fifty-six, and falling. Some country folk manipulated it skilfully; one report of 1830 noted that 'men of the poorer class in Derbyshire, drink large draughts of Foxglove tea, as a cheap means of obtaining the pleasures, or the forgetfulness, of intoxication'.

In the hands of doctors, it offered, and still offers, huge benefits, but at huge risk. It is yet another of the many Janus plants in this book: plants that look to healing in one direction and towards death in the other. But in this case it is not surprising that a plant that profoundly affects the functioning of the human heart can so easily kill the whole animal.

The two children on this wrapper (*c.* 1900) for healthy meat extract are being instructed about 'gift plants'. Presumably their teacher is mentioning the toxic possibilities of foxglove and other undesirable botanical 'gifts' as well as desirable ones.

THE FLIGHT
FROM PAIN

ALL HUMANS have suffered headaches, toothaches and migraines. All conditions have accumulated a large number of plants once used for their relief. The ancient Greeks tried powdered rose petals in oil. If that failed they tried a substance called galbanum – probably the dried rootsap from a relative of the fennel called *Ferula gummosa*. Dioscorides suggested an extract from *Physalis alkekengi*, now familiarly called Chinese lanterns. Pre-Columbian Indians in South America used another species of the same genus for aches and pains, and for a huge number of other ailments. (All species of *Physalis* are packed with exotic chemicals, many of which are currently under intensive investigation.) Mexican Indians tried an extract from the gorgeous perfumed flowers of a night-blooming cactus called *Selenicereus grandiflorus*, and may have got high into the bargain. Chinese and Japanese sufferers tried, as they did for almost every other ailment, ginkgo. More dangerously, they tried, and still use, an *Ephedra* species called ma-huan. They took the poisonous juice from the leaves of *Datura metel*, full of the witches' brew of atropine, hyoscyamine and scopolamine, or tried the poisonous extract from the scented yellow flowers of *Rhododendron luteum*. Chinese doctors suggested also extracts from the poisonous mountain peony, which cured wet dreams too. Fatal poisonings are still occasionally reported from the maladministration of all four plants.

Medieval Arab doctors suggested lesser things like attar of roses in vinegar, or the oils crushed from cinnamon or cloves. If they lived in the Yemen, they offered the narcotic qat and their patients drifted away, all smiles. Eighteenth-century colonial Americans sipped hot mint tea.

Glass flasks of distillates, pots of dried material, sponges, votive offerings, boxes of pills and more line the shelves of this Italian apothecary's shop, painted on the walls of Villa Issogna in the fifteenth century. Only the ambiguous glance between seller and buyer hints at the nature of the sale.

Nineteenth-century Frenchmen swore by absinthe. Only the Irish seem to have gone as far as having a goddess devoted to toothache: Pibheaeg offered devotees good plants like tormentil and, following Greek, Roman and Ayurvedic precedents, feverfew and chamomile.

Feverfew (*Tanacetum parthenium*), in spite of its deeply unpleasant smell, seems to have been used to treat headaches in Europe at least as soon as written records began. It is native to eastern Mediterranean countries, the Middle East and the Caucasus. It has been so important to the afflicted that it was taken to America with the first settlers and is now ubiquitous in suitable places throughout that continent too. It has also been widely used to lessen fevers, deflect hysteria, stop insect bites itching and more. Much modern research has looked at its hoped-for ability to lessen or stop migraines. Pharmacologists say that it is very likely that the sesquiterpene lactones in it inhibit the histamine and hormones known as prostaglandins, released during the inflammatory process. This prevents the contraction of the blood vessels in the head that probably trigger migraine attacks. Such a process might also explain feverfew's use to reduce toothache and the itch of insect bites. Researchers usually use a water extract of fresh leaves. Little work has been done on commercially available processed and stored material; its users report variable results.

Chamomile (*Chamaemelum nobile*), too, was once widely believed to lessen headaches when taken as a tea. A good 'shadow' plant – that is, one in the shadowlands of pharmacology, for which there is no proof that it works – with an ancient history, it was used for a huge range of ailments. Ancient Greek, Roman and Indian doctors used it for headaches and disorders of the kidneys, liver and bladder. The ancient Egyptians thought it cured malaria. It was widely used throughout medieval Europe. Later, Culpeper thought it was good for the usual agues, sprains, jaundice and dropsy, and as a shampoo. It is indeed a good fungicide. More enjoyably, both he and John Parkinson, whose book *Theatrum Botanicum* appeared in 1640, suggest that bathing in a light chamomile tea 'removes weariness and eases pain to whatever part of the body it is employed'. Recently it has been used for eczema and as a mouthwash to treat sores caused by certain sorts of chemotherapy, and to treat skin inflammation following radiation therapy. Experiments have cast extreme doubt on the first use, and completely exploded the second and third. Chamomile also contains coumarin compounds that can act as blood thinners, so its use is not advised if the patient is already taking something with the same effect.

But if feverfew, physalis, chamomile, galbanum or jaborandi have dubious effects on migraines and headaches, Willow trees (*Salix* spp.) have produced something astonishing. They given rise to a drug now so commonplace that we hardly think of it. Yet it has a history older than feverfew and the rest that stretches right back to ancient Babylon, and beyond. It also seems to be still packed with potential for the future. It is aspirin.

Its modern history started on 10 August 1897, when Felix Hoffmann, a chemist working for a German dyestuffs company called Bayer, managed to acetylate the phenol group of a

Chamomile, in an eighteenth-century illustration by Pierre Jean François for *Flore Medicale*. The plant has an ancient history of use in Egypt, India and Europe, for a number of conditions including headaches.

widely prescribed compound called salicylic acid. Hoffmann's new substance was acetylsalicylic acid. He was extremely excited. Salicylic acid produced from natural materials had been known since the 1830s. It was used, at high doses, to treat the pain and swelling in diseases such as arthritis, and to treat fever in illnesses like influenza. It had a bad side effect: acute stomach irritation. It did indeed reduce inflammation, but many sufferers found its effects too extreme to continue. Felix Hoffmann's father had arthritis and also suffered the side effects of salicylic acid. Felix thought that if he could alter the chemistry of the basic salicylic acid by incorporating an acetyl group into the molecule, the new formula might be less of an irritant. It was. The head of Bayer's pharmacology laboratory, Heinrich Dreser, was impressed. He even tested Hoffmann's new drug on himself. He then demonstrated the anti-inflammatory and painkilling effects of acetylsalicylic acid on animals. The new substance, given the commercial name of aspirin, was soon in use worldwide. Hoffmann's discovery paved the way for the modern pharmaceuticals industry.

Aspirin was only two steps removed from a living plant. The first step was made in 1829, when a pharmacist called Henri Leroux discovered a crystalline compound in the bark of willow trees that he called salicylin. This, he discovered, was the active chemical. A few years later, an Italian chemist called Raffaele Piria split up 'salicylin' to obtain salicylic acid in a pure state. This was much easier to administer than either natural bark extract or salicylin, and could be easily created in the laboratory or the factory. Felix Hoffmann made the third step.

The step before the chemists' first step was made thousands of years earlier. Clay tablets from the Sumerian period in Mesopotamia, one of the earliest of all urban cultures, describe the use of willow leaves to treat rheumatism. The ancient Egyptians also used the willow leaf and bark potions for their painkilling effects. Chinese doctors were using willow by 500 BC. Hippocrates prescribed it for women in painful labour; he also used it for aches and fever. In North America, the black willow (*Salix nigra*), which has a slightly different form of salicylin, was used by native Americans as a remedy for asthma, colds, influenza and indigestion, and as a diaphoretic and sedative.

In 1758, the Reverend Edward Stone (1702–68) of Chipping Norton in Oxfordshire began the first known experiments on willow bark. He ground up the drug and gave it to over fifty of his parishioners with ague – 1g of powder in half a glass of water every four hours. They all said their pain lessened, though the ague wasn't cured. He reported the experiment to the Earl of Macclesfield, president of the Royal Society in London. The society published his researches. Oddly, his interest in the willow was influenced not especially by the past but by his belief, almost exploded by the late eighteenth century, in a branch of the doctrine of signatures. He wrote: 'As this tree delights in a moist or wet soil, where agues chiefly abound, the general maxim that many natural maladies carry their cures along with them or that their remedies lie not far from their causes was so very apposite to this particular case that I could not help applying it; and that this might be the intention of Providence here, I must own, had some little weight with me.'

Earlier, European pharmacists had found similar effects in the bark of the white poplar (*Populus alba*). Both Gerard and Culpeper mention this as a cure for sciatica or 'ache in the hucklebones' (ankles), and the 'stranguary' (a result of a bladder infection). Gerard found it good against all inflammations and swellings, while Culpeper added that it dried up the milk in women's breasts after childbirth. Another plant that had been discovered to have similar painkilling properties was the delightful meadowsweet (*Filipendula ulmaria*) that so pleasantly perfumes the hedgerows and ditches of late summer on warm afternoons. Like willow and poplar, meadowsweet has bitter bark and was used as a mild painkiller. It too turned out to contain salicylin. Meadowsweet was, in the nineteenth century, put in the genus *Spiraea*, and Hoffmann seems to have used that genus to give the central syllable of the word 'aspirin'.

It wasn't until the 1970s that the mechanism for aspirin's effect was discovered. John Vane, who received both a Nobel prize and a knighthood for his work, found that it suppresses the production of prostaglandins. Though these hormones have many functions, they accumulate in damaged or infected tissues, and cause inflammation and the associated pain. Aspirin has other important effects. That it thins the blood had been observed since 1950, the initial work having been done by Dr Lawrence Craven, but in 1989 the mechanism was discovered. Prostaglandins also regulate the aggregation of the blood platelets that help to form clots. Suppress the prostaglandins, and the formation of clots that trigger heart attacks and strokes is also suppressed. A 1994 analysis of many experiments using aspirin showed that if people under seventy who are at risk of heart disease were to take aspirin regularly, the number of deaths from such disease across the world could be reduced by 100,000 a year. More recently still, it is becoming apparent that aspirin, again acting on prostaglandins, interferes with the biochemical mechanism that sometimes causes cells lining the bowel to become cancerous. Some researchers also believe that it may help treat or prevent dementia, high blood pressure in pregnancy, diabetes, gallstones, cataracts and migraines. These results are all from experiments using the double blind test. Approximately 100 billion aspirin tablets are taken each year. It looks as if this figure will increase.

Pain is so ubiquitous, and so feared, that its alleviation is easily exploited by the unscrupulous, the cynical and sometimes even the marvellously funny. One nineteenth-century quack, Perry Davis, said of his fabled 'Pain-Killer':

How much suffering could be prevented by a little foresight. Accidents happen, and sickness comes to ALL. Yet very many people never think of providing themselves with the means to promptly alleviate the sufferings from either.

An inexpensive and THOROUGHLY RELIABLE safeguard is offered in PERRY DAVIS' PAIN-KILLER which, for 49 YEARS has stood UNRIVALLED as a HOUSEHOLD REMEDY and TRAVELLING COMPANION. It is used externally as well as internally, and is just what is needed for BURNS, BRUISES, CUTS, SPRAINS &c.; and most people KNOW that NO OTHER remedy is to be compared with it as a CURE for COUGHS, COLDS, RHEUMATISM, NEURALGIA, &c., in Winter, and ALL SUMMER COMPLAINTS in their season.

Protect yourselves with a bottle at once, and be sure you have it with you when travelling.

IT IS A MEDICINE CHEST IN ITSELF.

Later slogans widened its appeal still further by proclaiming that it was 'Good For Man or Beast'.

Davis was born in poverty at Dartmouth, Massachusetts in 1791. Crippled at an early age, he was apprenticed as a shoemaker at fourteen. When he died, his name was known worldwide and he was very prosperous indeed. He was made rich by human gullibility. As a shoemaker, he eked out a living for his family, moving from town to town. He arrived in one, called Pawtucket, in 1828. He seems to have had an inventive bent, and sank his savings into the promotion of an improved design of grain mill. However, his scheme failed, and aged forty-seven he fell ill.

> I told my wife [he later wrote] that she could not expect to have me with her much longer. A cold settled on my lungs. A hard cough ensued, with pains in my side. My stomach soon became sore, my digestive organs became weak, consequently my appetite failed; my kidneys had become affected. The canker in my mouth became troublesome … I searched the globe in my mind's eye for a cure during my illness and selected the choicest gums and healing herbs. These were carefully compounded creating a medicine to soothe the nerves and a balm to heal the body. I commenced using my new discovered medicine with no hope other than handing me gently to the grave.

Necessity being the mother of invention, and he being already many thousands of dollars in debt, Davis raised twenty-four dollars by selling his – and presumably his family's – horse, wagon and harness. With the proceeds, he made up a first batch of his 'choicest gums and healing herbs'. Cured of his near-terminal ailments, he tried selling his medicine in the markets around Boston. However, he wasn't yet a showman and most of it he had to give away. Finally, in the summer of 1843 he carried yet another batch of medicine to sell at the Pawtucket fair, which drew the curious from all over New England. While Davis was crying his wares, one brave soul, probably a 'plant', stood up, claiming a terrible pain in the stomach. Downing the medicine, he declared instant relief. That made the rest of the crowd want a bottle too. A year later, Davis was making his medicine on an industrial scale.

He made some serious converts. When an outbreak of cholera occurred in Asia, every Baptist minister setting sail from America was laden with cases of 'Pain-Killer'. During the American Civil War, even the United States government's horses were dosed with it. When Perry died in 1862, the company was taken over by his son, and later, his grandson. By the 1880s, its main offices were at 380 St Paul Street, Montreal, Canada, and 17 Southampton Row, Holborn, London. The advertising was always good. His bottles are now collectors' items. Some of his pamphlets were splendid. One, in colour, shows a paterfamilias, holding up a bottle of product and saying 'Good-By Doctor!' as the doctor leaves, a black servant holding open the door. The copy is filled with testimonials.

Cynical souls regarded the 'Pain-Killer' as a splendid joke. 'It was not right to give the cat the "Pain-Killer"; I realize it now. I would not repeat it in these days,' wrote Mark Twain in his *Autobiography*. 'But in those "Tom Sawyer" days it was a great and sincere satisfaction to

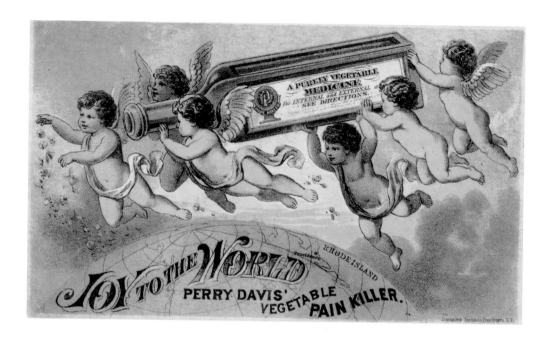

The half-dozen innocent cherubs shown on the trade card above may be scattering flowers and holding aloft the bottle but they probably had surprising names, amongst them 'alcohol', 'opium', and 'cocaine'. The formula for this hugely popular remedy was neither revealed nor discovered, but that didn't stop huge numbers of buyers from consuming it. Testimonials claimed that its 'choicest gums and healing herbs' cured everything from neuralgia to cancer.

me to see Peter perform under its influence – and if actions do speak as loud as words, he took as much interest in it as I did. It was a most detestable medicine, Perry Davis Pain-Killer. Mr. Pavey's negro man, who was a person of good judgment and considerable curiosity, wanted to sample it and I let him. It was his opinion that it was made of hell-fire.'

The formula for the 'Pain-Killer' seems to be unknown. Many such quack medicines, especially ones suggested for internal as well as external use, contained huge amounts of alcohol. Some were laced with coca leaves, various opiates, even terrible poisons such as nux vomica. They were as popular, and as profitable, in Europe as they were in North America. The huge success of some of these quack, or partially quack, medicines was made possible by the advent of cheap print, and the wide dissemination of newspapers, journals and magazines. London newspapers of the late nineteenth century were packed with adverts. Readers were assured: 'Freedom from Cough in Ten Minutes after use, is insured [sic] by Dr LOCOCK's PULMONIC WAFERS'. Or if the stresses of life began to tell, all they had to do was buy 'Norton's CAMOMILE PILLS for indigestion, bilious and liver complaints, headaches, heartburns, acidity etc'. Indigestion must have been a constant problem, for every publication was full of remedies, bona fide and otherwise. Only a few were effective, or survived throughout the century in the way that Milk of Magnesia did.

The urban poor, without access to garden ground, were completely at the mercy of quack and semi-quack. They could, of course, try to buy herbs from the markets. In 1871, one doctor working at the Manchester Infirmary reported on what he saw at the large herb markets of the city. He came to the conclusion that there were three levels of quacks. The worst were those using undisclosed *materia medica*. They wrapped their cures in mumbo jumbo, and preyed on the rich or the desperate. Another rank were much closer to proper herbalists, using old herbals and consulting real doctors. They worked largely for or on the middle class. Lastly, the doctor saw what he called 'old crones and their believers', often working for little or no reward, and handing out folk remedies to the very poor. In the Manchester markets, trade was brisk. Vast amounts of medicinal herbs were being sold, often with the sellers not knowing the Latin name, or even the local common name. Plants such as aconite, digitalis, comfrey and chamomile were commonly sold by handful or armful with very little instruction about dosage. The poor simply had to take their chance in this botanical roulette, for they couldn't afford to go to a qualified apothecary.

The new wave of cheap and popular newsprint had its good side too, for it enabled real help for the poor as well as their exploitation. In the 1850s, in one of Britain's new gardening newspapers, a retired doctor wrote a series of articles for the poor, using the contents of his own garden as a starting point. He wrote, modestly, that but a page would

> ... *describe its contents, and their application in disease and sickness, but yet it is abundantly useful to my neighbours. Nay, the village doctor himself sometimes borrows from my beds, and, though somewhat jealous of what he calls my unprofessional conduct, we are, upon the whole, mighty good friends; for to tell the truth, he somewhat leans upon me ... but here I must premise that to cull simples and to prepare and administer their products with success, require both skill and experience. It is a task well suited to*

a physician like myself, who can afford to sit down under the tree of his old age, and devote himself to such a speciality.

But he was interested mostly in the mild 'simples' of country gardens. He offered no help, at least in print, for anyone in really serious and terrible pain. The use of Janus plants such as henbane and aconite, as we have seen, requires the sort of knowledge and skill impossible to impart in such a diffuse way.

The most efficient way to use a skilful practitioner has always been to create hospitals around them. The 'hospital' idea, of places where the needy and the highly trained physicians, surgeons and apothecaries needed to help them could all be concentrated, seems to have emerged slowly. Once started, it spread fast. A great institution in Baghdad was founded by Haroun al-Rashid in AD 805. The West followed suit. There were many hospitals in England's southern counties by 1150 and further north, in impoverished states such as Scotland, major hospitals were founded soon after.

The modern traveller by road has many options for heading south from the ancient Scottish capital, Edinburgh. One road, heading due south and meeting the bleak Lammermuir Hills head on, crosses the River Esk at Dalkeith, then gradually climbs upwards through Pathhead, past the ancient village of Fala. There, the traveller sees the grey, windswept top of Soutra Hill. In winter, the road is often closed by snow. In summer, there is often slow traffic. At the base of the hill, where the road swings up and eastwards, the impatient motorist will probably not notice a small side road that turns westwards to approach the heights, or a small yellow sign saying 'Soutra Aisle'. The hillside is bleak and open, and without buildings. Yet if our traveller were swept back in time to the year 1462, his horse, having splashed through the Dean Burn and stumbled up the ravine side, would have afforded its rider an astonishing view.

Soutra Hill is capped by an immense walled citadel; above the wallhead rises a great church tower, the gables and chimneys of many handsome buildings. Around the walls cluster stables, outbuildings, the hovels of servants and grooms, fortune-tellers' booths, pigsties, hen houses, rubbish pits and middens. The Roman road cuts its way through them all on its way, ultimately, to Rome. The Augustinian hospital at Soutra, this year at its apogee, and already 354 years old – its first charter was dated 1108 – is endowed with huge estates. It can welcome travellers or searchers of sanctuary. It can accept the poor, the sick and the wounded, sometimes accommodate whole armies and their royal master, whether Scottish or English. For the wants of all, it contains within its walls three sheltered gardens devoted to growing medicinal plants. The infirmarer's shelves are piled with green-brown earthenware albarello jars containing salves, ointments and extracts from the plants in the garden and from the estates. They also hold substances that have travelled the great Silk Road, whose branches radiate into the deserts and harbours of Somalia and Yemen, the shores of southern India, and China.

For Soutra, it is all about to end. In Edinburgh, the widowed Queen Mary of Gueldres has founded Trinity Collegiate Church. It is to have an associated hospital near her tower house at Holyrood. To finance her grandiose scheme, she has this year annexed the estates and revenues of Soutra. The hilltop hospital, the most advanced and famous in the north, will slowly decline in importance until in the sixteenth century it is extinguished.

Though the hospital was completely razed by the late 1800s, its drains, cesspits, soakaways and middens still exist and excavations of these began in the 1980s. They are packed with grisly remains: huge amounts of coagulated blood, bones of amputated parts, eggs of parasites and spores of the dread disease anthrax. There are also the remains of plants: pollen, seed, fragments of leaves and stems. The hospital flora contains around two hundred species, from the opium poppy, hemp, tormentil and valerian that were grown locally, to the myrrh and galingale that had travelled so far.

The association of some plants with certain sorts of hospital waste provides clues to their use. Though tormentil was also used as a painkiller, here its fragments were found with the eggs of several sorts of ascarid worm. It was probably used to try to staunch the bloody diarrhoea caused once the numbers of parasites had built up to huge numbers in the sufferer's intestine. Another sort of waste, associated with the bones of amputated limbs, contains a terrifying triad of Janus plants: henbane, aconite and opium poppy. The three occasionally appear in medieval written sources as a combination used as a painkiller during radical surgery. Henbane and aconite are both exceptionally poisonous. The very greatest care would have been needed in the making up of the mixture. The apothecaries at Soutra must have been very highly skilled.

Henbane (*Hyoscyamus niger*), like so many medicinal plants that actually react with the human body, has, like digitalis, a low therapeutic ratio. But though the difference in dosage between the plant doing something useful and killing the patient is extremely narrow, in truly minute doses it has some remarkable actions.

Native to Europe, western Asia and North Africa, henbane is certainly a strange-looking plant. A rosette of pallid fleshy leaves throws out long leafy stalks bearing drab but sinister purple and grey spotted flowers. Only the mandrake and belladonna look as alarming. Dead wanderers along the banks of the Styx, Hades' river boundary, wore garlands of henbane in their hair, perhaps as a warning to the living to take the plant very seriously indeed.

It was certainly in use even long before it was particularly recommended by Dioscorides in the first century AD. He used it, benevolently, to let his clients sleep and mitigate their pain. He and his contemporary, Celsus, used it pounded into an oily salve that was rubbed on to the skin. It was much safer so administered, and also acted as a local anaesthetic. Taken orally, the slightest overdose resulted in a stupor that Pliny thought to be 'of the nature of wine and therefore offensive to the understanding'. A little stronger still, henbane was used in the cults of the dark gods and, once Christianity took hold, driving the older religions underground, in the rites of magicians and diabolists. Even witches treated it with caution. Pounding it with bear's grease, they smeared the resulting mix on their bodies. They and their broomsticks took flight, the henbane letting them career amongst the stars, locked in ecstatic embrace with their strange familiars. Even the dried roots were potent. Amulets and necklaces made of them were given to children. They were supposed to suppress fits, and the frets caused by the appearance of new teeth. The children were

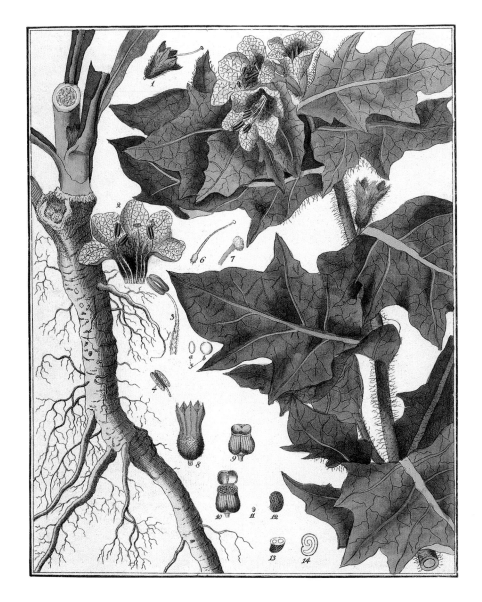

Paracelsus mixed the juices of henbane (*Hyoscyamus niger*) with crushed pearls, frogspawn and opium to make his famous cure-all 'laudanum'. Despite its association with witchcraft, henbane was extensively grown by the Shakers in their Physic Garden at New Lebanon, New York. This coloured engraving by Peter Haas was published in 1805.

probably absorbing small amounts of its alkaloids of scopolamine and hyoscyamine through their skins.

Smoke from charring seeds was sometimes funnelled into the mouth to relieve the agony of toothache – much cheaper, if far more dangerous, than oil of cloves. Sometimes steam from boiling water containing seeds was used in the same way. Some of its properties were

With broomsticks between their thighs, liberally greased with salve, these witches – in an engraving of 1579 by an English artist – take off dangerously up the sort of chimney that allows Santa Claus down. Their activities, henbane-based and unlawful, are doomed by the spying onlooker.

so strange that by the time the Soutra hospital was falling down, henbane was in use by medical mountebanks. Gerard noted that 'Drawers of teeth who run about the country and pretend they cause worms to come forth from the teeth by burning the seed in a chafing dish of coals, the party holding his mouth over the fume thereof, do have some crafty companions who convey small lute strings into the water, persuading the patient that these little creepers came out of his mouth, or other parts which it was intended to ease.' The trick was even easier than Gerard thought: the small white, cylindrical embryos inside the seed are forced out by the mounting pressure caused by the heat and the mountebanks duped the ignorant sufferers into believing that the embryos were 'worms' from their teeth. By the time the toothache returned in a few hours, the conmen were out of range of repayment.

Culpeper, trying to refute earlier authors, wrote: 'the herb is indeed under the dominion of Saturn and I prove it by this argument: All the herbs which delight most to grow in saturnine places are saturnine herbs. Both Henbane delights most to grow in saturnine places, and whole cart loads of it may be found near the places where they empty the common Jakes, and scarce a ditch to be found without it growing by it. Ergo, it is a herb of Saturn.' He may not have been aware of mountebanks' tricks, for he writes later: 'The oil of the seed is helpful for deafness, noise and worms in the ears, being dropped therein.'

Not surprisingly, such a peculiar and useful herb travelled swiftly across the Atlantic. It seems to have started off in New England, and by the 1670s was established in all European settlements in North America. Rather than Black Sabbaths, or even toothache, it seems to have been associated with cattle rearing, for a wash made with either the leaves or the seed killed lice on man and the appalling ticks and maggots that fasten on to cattle. It reached Brazil in the same century too, where again it was used as a cattle drench. Back in New York, the 50-acre Physic Garden of the Shakers at New Lebanon was producing, by 1850, substantial quantities of henbane, aconite and opium poppy, just as the Soutra hospital gardens did seven hundred years earlier.

At Soutra, where there were probably no witches, though numbers were burnt in nearby villages into the eighteenth century, the henbane was probably used only as Dioscorides used it: to cause deep sleep. Oddly, the waker was supposed to retain no remembrance of the immediate past – helpful if undergoing the trauma of major surgery. It seems likely that those who needed radical cutting at Soutra were drugged with henbane first and then with the other two in the triumvirate of potent plants. If the concoction worked, it seems strange that it later dropped out of use. Perhaps it was simply too dangerous.

If henbane was bad, aconite was worse. The word refers not to the small yellow flower of early spring, *Eranthis hyemalis*, widely grown in our gardens. That is also poisonous, but it is as nothing compared to almost all the species of *Aconitum*. Variously called monkshood, friar's cap, auld wife's huid, as well as aconite, the hooded flowers are in shades of greenish or purplish blues, sometimes white, sometimes livid yellow, rarely a dingy pink. Many are of great beauty, though all are so poisonous that they should be planted in the garden only after due consideration. The species used medicinally in Europe is *Aconitum napellus*, the specific name referring to its turnip-like root.

Some species of aconite were well known to the ancients as deadly poisons. The plant was supposed to be the invention of Hecate, who took it from the foaming mouth of Cerberus, the terrible dog that guarded the entrance to Hades. It may have had social uses: for instance, a species of aconite was put in the drink that the old men of the island of Chios took when they became infirm and no longer of use to the community. Medea is supposed to have used aconite when she was trying to poison Theseus.

Probably not native to Britain, *Aconitum napellus* is found in English lists of plants from the tenth century onwards. In the Anglo-Saxon, it is called thung, though this was also a general name for any very poisonous plant. One of its old names is also wolf's bane, the direct translation of the Greek *lykŏs ktŏnŏs*. Greek shepherds, and others, may indeed have used it to kill wolves, bait dead lambs with the juice and perhaps use it as an arrow poison. By Shakespeare's day, helmet-flower was its ordinary name. The older herbalists describe it as venomous and deadly. Gerard writes: 'There hath beene little heretofore set down concerning the virtues of the Aconite, but much might be saide of the hurts that have come thereby … the herb only thrown before the scorpion or any other venomous beast, causeth them to be without force or strength to hurt, insomuch that they cannot moove or stirre untill the herbe be taken away.' There were no scorpions in London in Gerard's day, so it isn't clear from where he got his information. Ben Jonson, in his tragedy *Sejanus*, says:

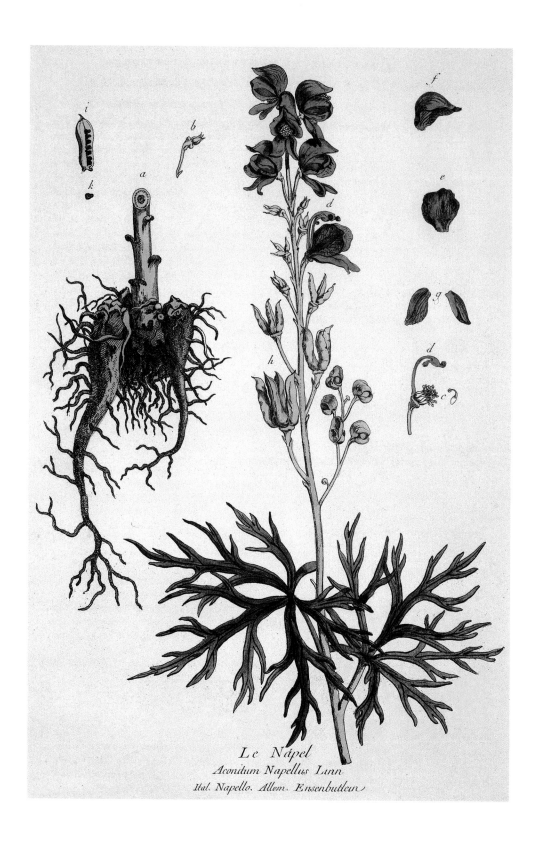

Le Nápel
Aconitum Napellus Linn
Ital. Napello. Allem. Ensenbutlein

LEFT The ancients believed that the deadly aconite was an invention of the goddess Hecate, made from the mouth foam of the three-headed dog Cerberus. Even though it is more dangerous than cyanide, it found wide use in nineteenth-century medicine. This illustration by F. and G. Renault appeared in *La Botanique, Mise a la Portée de Tout le Monde* (1774).

BELOW Giambattista della Porta, a free-thinking radical interested in the stranger sides of ancient Roman medicine, was drawn to alchemy and witchcraft. He was hounded by the Inquisition.

I have heard that Aconite
Being timely taken hath a healing might
Against the scorpion's stroke.

Certainly great use was made, at least in Italy, of its deadlier attributes. In 1589, an aristocratic Neapolitan called Giambattista della Porta published the second edition of his *Magia Naturalis*, in which he describes various methods of poisoning. He gives a formula for 'Veninum Lupinum', which was a nasty mix of aconite, the common yew (*Taxus baccata*), caustic lime, arsenic, bitter almonds and powdered glass. Mixed with honey, it was made into pills the size of walnuts. Neither size nor taste could have appealed to the potential victims. Wolves or humans, had they tried a bite, they would have felt tingling and numbness of tongue and mouth, a sensation of ants crawling over skin gone cold and clammy, as well as suffering nausea and vomiting with laboured breathing, an irregular pulse, giddiness and staggering. Their minds would have remained clear. For humans, a nineteenth-century manual suggests, as a cure for poisoning by yew: 'A stomach tube or emetic should be used at once, 20 minims of Tincture of Digitalis given if available, stimulants should be given and if not retained diluted brandy injected per rectum, artificial respiration and friction, patient to be kept lying down.' Della Porta eventually got into grave trouble with the Inquisition, for he was master of a secret academy studying witchcraft, natural magic and more. In the 1560s he had experimented with the 'witch's salve', or sorcerers' pomade, supposedly used to transport witches into flight. Della Porta maintained that its effect could be understood according to scientific principles. He believed that it was nothing more than a sleep-inducing hallucinogenic drug, which when rubbed on the body caused supposed witches to fantasize their nocturnal flights. Even that still troubled the Church.

In della Porta's Veninum Lupinum mix, one-fiftieth grain of the plant's main toxin, aconitine, would kill a sparrow in a few seconds. Well into recent times, the only way of standardizing the aconitine content of any preparation was to find out how much was needed to kill a guinea pig. Even ointments and tinctures made from it were dangerous. Gardeners who like the plants should wear gloves when handling them. Any juice that gets into a wounded finger affects the whole body, which will feel not only pains in the limbs, but a sense of suffocation and syncope.

The genus contains about one hundred species. Northern India and China are especially rich. Each species has a slightly different mix of alkaloids and a varying amount of aconitine itself. Many species are wild collected and exported to the West. In China, the handsome autumn-flowering *A. carmichaelii* is the most used. It is frequently combined with cinnamon in various medicines that balance ying and yang in the right-hand kidney. Most cases of serious poisoning reported in Hong Kong are related to *A. carmichaelii* and another species, *A. kusnezoffii*. Both are also used to treat rheumatism, arthritis, bruises and fractures. One recipe for you gui wan ('Restore the Right Kidney Pill') contains, amongst other ingredients, Chinese foxglove (*Rehmannia glutinosa*), aconite, yam (*Dioscorea* spp.) and Chinese dodder (*Cuscuta chinensis*). All are packed with active ingredients.

In nineteenth-century Europe and North America, aconite was, like so many dreadfully poisonous plants, immensely popular with doctors. They used it internally as a painkiller, a diuretic and a diaphoretic. It was employed to lessen the rate and force of the pulse in the early stages of fevers, and for laryngitis, as well as the first stages of pneumonia and erysipelas. It also relieved the pain of pleurisy and aneurism. As a liniment for external use, mixed with chloroform or belladonna, it did all the things it did at Soutra hospital, as well as diminish the pain of neuralgia, lumbago and rheumatism. The Shakers grew large quantities of it at the Physic Garden at New Lebanon, New York. It is much used *in absentia* – that is, in such extreme dilutions that it is absent – in homoeopathic remedies.

The writer and courtier Fanny Burney (1752–1840), seen here in an engraving of 1842. She underwent a major operation without the use of any sort of painkiller, even though opium and other narcotics were widely available. Substances such as chloroform did not appear until 1847.

Although the painkilling triumvirate of henbane, aconite and opium poppy appears to have been used at Soutra, and in many other medieval hospitals, something caused surgeons to turn their backs on it. By the eighteenth and early nineteenth centuries, and until the gradual popularization of chloroform in the mid-nineteenth century, many people needing surgery had no help with pain whatever. Their agony can hardly have helped their survival. Here is a fragment of the courtier, novelist and diarist Fanny Burney's appalling account of a mastectomy operation carried out on her in 1811:

M. Dubois placed me upon the Mattress, & spread a cambric handkerchief upon my face. It was transparent, however, & I saw through it that the Bed stead was instantly surrounded by the 7 men and my nurse, I refused to be held; but when, bright through the cambric, I saw the glitter of polished steel – I closed my eyes …

Yet – when the dreadful steel was plunged into the breast cutting through veins – arteries – flesh – nerves – I needed no injunctions not to restrain my cries. I began a scream that lasted uninterruptedly during the whole time of the incision – & I almost marvel that it rings not in my Ears still! so excruciating was the agony. When the wound was made, & the instrument was withdrawn, the pain seemed undiminished, for the air that suddenly rushed into those delicate parts felt like a mass of minute but sharp & forked poniards, that were tearing the edges of the wound, but when again I felt the instrument – describing a curve – cutting against the grain, if I may so say, while the flesh resisted in a manner so forcible as to oppose & tire the hand of the operator, who was forced to change from the right to the left – then, indeed, I thought I must have expired, I attempted no more to open my eyes … The instrument this second time withdrawn, I concluded the operation over – Oh no! presently the terrible cutting was renewed – & worse than ever, to separate the bottom, the foundation of this dreadful gland from the parts to which it adhered … yet again all was not over.

Her account goes on for another few pages, becoming even more dreadful. Perhaps because of her court connections, at least the knives were clean; and she survived until 1840. The pain was clearly terrible, and though she might not have wanted the full witches' triumvirate of henbane, aconite and opium, even straight opium would have helped her. There was plenty of it around then, but it was not being used for pain relief.

Humans and the opium poppy (*Papaver somniferum*) have had a long and complicated relationship. The poppy seems to have come into domestication, if such a word can be used for such an extraordinary plant, at the same time as various grasses and other seed crops began to be utilized. Perhaps it started life as a seed crop; the seed is abundant and delicious, and yields copious oil. The seed also contains a slight amount of narcotic substances and, eaten in quantity, fresh, allows a pleasant relaxation. Perhaps it was that which set in train more thorough research into the plant's properties. In the Middle East, Neanderthal man may have used the opium poppy in sites dated 30,000 years old. Once writing began, its history is clearer. By around 3400 BC, the Sumerians referred to it as Hul Gil, the 'joy plant'. They exported that joy to the Assyrians, who seem to have been aware that the pods can be scored with a knife to yield a strongly narcotic latex. The

Babylonians copied the Assyrians, and the Egyptians copied the Babylonians. The plant was in northern Europe by 2000 BC; remains of poppy-seed cake and poppy pods have been found in Swiss lake dwellings. By 1300 BC, the Egyptians grew a very special sort of poppy, perhaps a particularly potent variety, at Thebes. Its product was called *opium thebaicum*, and its trade vastly enriched the pharaohs Thutmose IV and Akhenaton, and provided substantially for the golden King Tutankhamen. From Thebes, its trade stretched out along the routes sailed by the Phoenicians and Minoans. Everywhere, the poppy was taken up. Two hundred years later, superb-quality scoring knives, used to scar the young seed pods, were being made in Cyprus.

By now, too, the latex that oozed from the pods was being macerated in wine. Since alcohol dissolves a wider range of potent alkaloids than does mere water, the maceration produced different effects. Homer conveys its effects in the *Odyssey*. In one episode, Telemachus is depressed after failing to find his father Odysseus. But then Helen 'had a happy thought. Into the bowl in which their wine was mixed, she slipped a drug that had the power of robbing grief and anger of their sting and banishing all painful memories. No one who swallowed this dissolved in their wine could shed a single tear that day, even for the death of his mother or father, or if they put his brother or his own son to the sword and he were there to see it done.'

By about 460 BC, Hippocrates, a rationalist, was trying to clear away the magical properties that were rapidly accumulating around the drug so that physicians could concentrate on what it could actually do. A century later, Alexander the Great's armies introduced opium to the peoples of Persia and India. Galen listed its medical indications, noting how opium 'resists poison and venomous bites, cures chronic headache, vertigo, deafness, epilepsy, apoplexy, dimness of sight, loss of voice, asthma, coughs of all kinds, spitting of blood, tightness of breath, colic, the lilac poison, jaundice, hardness of the spleen stone, urinary complaints, fever, dropsies, leprosies, the trouble to which women are subject, melancholy and all pestilences.'

Opium could readily be bought on the street markets of ancient Rome. By the eighth century AD, opium use had spread to Arabia, India and China. The Arabs both used opium and organized its trade. The Prophet prohibited the use of alcohol, but not hashish or opiates. Indeed, by 400, Arab traders were exporting *opium thebaicum* as far afield as China, where it was to have, over the next 1,500 years, a devastating effect. By 1200, opium was penetrating all the medical schools of India, but while the strange purple flower gradually colonized further and further east, for some reason, in the West, it vanished from the literature. Perhaps, as at Soutra in Scotland, it survived. There is evidence for its cultivation in the coastal farms belonging to the hospital. Nineteenth-century experiments showed that even a Scottish summer can ensure a reasonable amount of opiates in the pods' sap. Perhaps

During its long association with mankind, the opium poppy has been selected for huge numbers of variants, some extremely decorative. This one with double flowers and frilled petals, illustrated in *Hortus Eystettensis*, can still be found. All contain opiates.

FAMOSO·DOCTOR PARESELSVS

Paracelsus, or Theophrastus Bombastus von Hohenheim, not only gave us the
word 'bombast' but also broke apart the ancient conventions of medical thought,
allowing new ideas to develop. This painting is a contemporary copy after the
lost original by Quentin Massys (1466–1530).

it went 'underground'. Some authors have blamed the church for making the plant taboo, so associated was it with the pagan Rome and the evil Moors.

However widely it may have been used in real life, in literature it had to wait for rehabilitation until the 1500s, and the extraordinary theories of an extraordinary man: Paracelsus. His real name was Theophrastus Bombastus von Hohenheim (1493?–1541). The nickname echoed his belief that he was superior to one of the greatest classical physicians, Celsus. Paracelsus was probably deranged, and his formidable, quarrelsome, violent self-inflation gave rise to the word 'bombast'. Born in Einsiedeln in Switzerland, Paracelsus received a degree in medicine, possibly from the University of Vienna. Like many doctors, he drew no division between that discipline and those of alchemy and astrology. He travelled to search out the hidden secrets of all three, and his theories became more and more wild. He broke completely with contemporary theories, which were descended from the writings of Galen and Hippocrates. He asserted that diseases were caused by agents that were external to the body, which could be countered by chemical substances. He meant alchemical ones, but was the first to use sulphur and mercury compounds against syphilis. He was therefore responsible for the occasional curing and frequent poisoning of the promiscuous that continued well into the nineteenth century. Still, his revolt against ancient medical precepts freed medical thinking from the past, enabling it once more to take a more scientific course. What could have been more important for the likes of poor Fanny Burney was Paracelsus's introduction of what he called 'laudanum'. Having read the classical authors, he was aware that opium latex gave more freely of its contents when macerated in alcohol, dissolving out the compound now called morphine. In 1527 he started to play around with some sinister-looking black pills that, with his passion for inflation, he called the 'Stones of Immortality'. He soon proclaimed: 'I possess a secret remedy which I call laudanum and which is superior to all other heroic remedies.' He had extracted the black lumps of *opium thebaicum* into brandy, and added exotic extras such as crushed pearls, tincture of henbane and frogspawn. Though he wrapped up the idea in alchemical mumbo jumbo, laudanum went on to have a long life. (Laudanum can also refer to the soporific juice that oozes from lettuce stems.)

Back in Britain, opium was also making converts. Elizabeth I instructed merchants to purchase the finest Indian opium and transport it back to England. The recipes for laudanum changed. Thomas Sydenham (1624–89), a London physician who fought as a captain in Cromwell's army during the English Civil War, standardized laudanum to the now classic formulation: 2 ounces of opium, 1 ounce of saffron, a drachm of cinnamon and cloves – all dissolved in a pint of Canary wine. By 1680, he was selling what he called 'Sydenham's Laudanum'. Not surprisingly, it became a popular remedy for numerous ailments. He was later called 'the English Hippocrates'.

By the time of Miss Burney's operation in 1811 vials of laudanum and raw opium were freely available at any English apothecary's or grocer's store, and both were soon to become even more popular. In 1800, the British Levant Company was buying up about half of all the thousands of cases of opium on the market at Smyrna (modern Izmir), Turkey. Grown in the grey dusty heartlands of Anatolia, this was the most potent of all opium then grown in the Middle or Near East. (Such high-quality material was strictly for importation to Europe and the United States. Weaker crops were sent off to China.) The British liking for medicinal and recreational use of opium meant that in 1830 the country imported 22,000 pounds. By 1860, that figure had multiplied ten times. Opium, as painkiller and narcotic, and whether with a water or alcohol base, was everywhere: in baby soothers, indigestion remedies and even travel sickness cures to use in the new railway carriages.

Gradually, however, the dark side of the poppy began to affect world events. The British East India Company, which traded extensively in India-grown opium, was entirely aware of opium's ability to enslave people. It was anxious to promote and preserve addiction in one of its most important markets: China. In that vast country, the need for opium amongst the populace was so strong that the country's wealth was slowly slipping away as its population spent more and more on the imported drug. Opium was already part of the culture. A tenth-century poem related how to brew an opium drink 'fit for Buddha'. Drinking may have been the main method of consumption. The Chinese certainly knew about smoking opium by the sixteenth century, but considered the method too barbaric to use. This view changed in the seventeenth century, when tobacco smoking in pipes was introduced by the Dutch from Java. Chinese devotees discovered that Indian opium mixed with tobacco gave an extremely potent smoke, the effects of which were instantaneous. The opiates that didn't turn to vapour congealed as 'dross' in the pipe bowl, so even old choked-up opium pipes were valuable: the dross could be harvested and sold on to the poor.

By the late 1700s, the East India Company controlled not only the prime Indian poppy-growing areas between Patna and Benares but also the distribution and sale of the product. Opium created huge wealth for the company and its directors, many of whom built themselves splendid classical mansions surrounded by green acres of the English shires. The company held highly profitable auctions of the right to sell opium in various sections of its captive market. To save China from ruin, the Imperial court banned its use and importation. Smuggling began at once. In 1839, the Qing emperor Tao-kuang ordered his minister Lin Tse-hsu to take action. Lin petitioned the new British monarch, Queen Victoria, for help. She ignored his pleas. The Emperor had 20,000 barrels of opium and many of its British and Chinese dealers seized. The British retaliated by attacking the port city of Canton. The first of the deeply discreditable opium wars had begun.

Everyone wanted a slice of the opium trade. Various American entrepreneurs attempted to hive off part of the trade to China by supplying the Chinese with high-quality opium from Turkey. One of these was John Cushing, who, in the employ of his uncles' business, James and Thomas H. Perkins Company of Boston, was fabulously wealthy by 1812. Four years later, John Jacob Astor bought ten tons of Turkish opium to ship to Canton aboard the *Macedonian*. Later, he switched to supplying the English habit alone. Soon, many well-known people were using opium in a non-medicinal way, most at non-addictive intervals. Those who used it included John Keats and other writers. Samuel Taylor Coleridge wrote 'Kubla Khan' under its spell (opium promotes vivid dreams as well as gentle euphoria).

In Xanadu did Kubla Khan
A stately pleasure-dome decree
Where Alph, the sacred river, ran
Down to a sunless sea ...
Beware! Beware!
His flashing eyes, his floating hair!
Weave a circle round him thrice,
And close your eyes with holy dread,
For he on honey-dew hath fed,
And drunk the milk of Paradise.

While some never became addicted, others got hooked. Thomas De Quincey, who published his autobiographical account of opium addiction, *Confessions of an English Opium-eater*, in 1821, found the 'Divine Poppy-juice, as indispensable as breathing'. Baudelaire likened opium to a woman friend, 'an old and terrible friend, and, alas! like them all, full of caresses and deceptions'. Elizabeth Barrett Browning fell under its spell, though she found morphine did not threaten her ability to write 'poetical paragraphs'. In America the use of opium was no different. William Blair described his experiences with opium in a New York magazine of 1842:

... while I was sitting at tea, I felt a strange sensation, totally unlike any thing I had ever felt before; a gradual creeping thrill ... and before my entranced sight magnificent halls stretched out in endless succession with galley above gallery, while the roof was blazing with gems, like stars whose rays alone illumined the whole building, which was tinged with strange, gigantic figures, like the wild possessors of a lost globe ... I will not attempt farther to describe the magnificent vision which a little pill of 'brown gum' had conjured up from the realm of ideal being. No words that I can command would do justice to its Titianian splendour and immensity.

Thomas De Quincey (1785–1859) first met the poppy in 1804 at Oxford University. He used it to relieve acute neuralgia pains, and he lived with a decanter of laudanum permanently at his side. This engraving is by Francis Croll.

The Chinese populace became so enamoured of the cheap opium produced by the
British in India that smuggling became a prodigious enterprise. In this woodcut of *c.* 1850,
bales of the drug are being hauled over a city wall.

As a result of such addictive use, opium's image began to darken. Nevertheless, the
medicinal or quasi-medicinal usage of it spread widely. With its inclusion in remedies for
children, babies were introduced to the pleasures of opiates at their mothers' breast.
Godfrey's Cordial, a syrup of opium tincture effective against colic, was vastly successful.
Street's Infants' Quietness, Atkinson's Infants' Preservative and Mrs Winslow's Soothing
Syrup were all made to similar recipes. 'Dr Locock's Powders for all Disorders of Children,
including Chicken-pox, nettle rash, measles, scarlatina, sore eyes, wasting, rickets, etc. etc.'
was another. Children began to cry if they were not dosed regularly. As it had in great
Galen's day, opium now seemed to offer a cure for every malady of man. Husbands noticed
rising expenditure on family cure-alls, if they weren't by then hooked themselves.

Meanwhile the Chinese were inevitably defeated by the superior technology of the British
navy and the Emperor was forced to accept the Treaty of Nanking, which left his country
open again to the increasing economic devastation caused by his people's liking for opium.
China was forced by circumstance to make another attempt to control the poppy, and in
1856 the outraged British, aided by the French, started the second opium war. Again China
could not win, and was forced to pay another huge indemnity. The importation of opium

During the first opium war (1839–42), the British government supported the opium trade by force. Here the warships *Imogene* and *Andromache* are protecting the fleeing smugglers' junks and skiffs by attacking the Chinese fort.

was legalized, and, ironically, opium production began to take place along the highlands of South-East Asia and western China, all now in the area that we call the Golden Triangle.

In the West, the poppy was extending its reach. Chemists were looking at ways of refining opium further, and trying to discover its active ingredients. In 1843, an Edinburgh doctor, Dr Alexander Wood, began experimenting with the results of the chemists' work, administering morphine to patients by means of a hollow needle he'd soldered to the end of a small syringe. In 1895, a German dentist tried altering the structure of morphine in order to lessen some of its side effects. By adding an acetyl group he produced diacetylmorphine, a compound almost at once called heroin. It turned out to be even more addictive than morphine.

So the simplest of medicinal plants is most truly a Janus plant. So easily grown, so long with us, so beautiful in some of its garden forms, it has wrecked, is wrecking and will continue to wreck, millions of lives. It reaches deep into our own Janus natures. Yet whoever has experienced appalling pain cannot have helped but bless the plant that, after the quick jab of a needle, allows the pain to shuffle to a distant room. Anyone who has seen the terminal agonies of a loved one soothed, cannot help but thank the wreath of poppies that Death sometimes wears. It is indeed a most extraordinary plant.

CHASING VENUS

NIGHT. A grey undulating plateau; only starlight falls into the deep valleys that cut down its western edge towards the sea. In one great valley of darkness, there is the rustle of much movement, muffled bellowing, the rattle of slipping stones. The first of the freight camels sways up the trackway to breast the edge of the ridge. More dark forms sway upwards from the dark. The caravan is following a route already a thousand years old, the stones compacted, polished, glowing paler than the surrounding desert. From the mass of animals, a long-legged fighting camel is whipped faster onwards, breaks ranks and sets off to a higher ridge. Its harness glitters, golden, beaded. The air on the heights is cold. It is free of the smells of humans and camels, and of the contents of the bales and baskets: myrrh, frankincense, nutmeg, mace, ginger, cinnamon – only the gold and precious stones have no

fragrance. The warrior on the golden camel turns to look at the sea. She believes herself the earthly incarnation of the moon now risen. She is going to mate with the sun.

The route that the Queen of Sheba is following will continue in use for another thousand years. Before its fall, the Greek geographer Strabo (?64 BC–?AD 23) would compare its immense traffic to the appearance of an army in transit. Pliny the Elder (AD 23–79) will describe its sixty-five stages, each a suitable halt for the camels. Nearly two thousand years after that, it will be referred to as the Incense Road.

Myrrh and frankincense were the most valuable commodities transported along the route. The beads of aromatic resins, called 'tears', were harvested in the fertile lands of the wealthy theocracies along the southern shore of the Arabian peninsula and carried along narrow, scarcely marked tracks starting in Hadhramaut and Saba. These tracks finally braided together into a great route that ran westwards along the coast, swung up through the mountains that nearly reach the sea at the Gulf of Aden, then went up the western seaboard

Every age has its own Queen of Sheba. Here, on the side of a cassone painted by Apollonio di Giovanni *c.* 1450, she is seen through Italian Renaissance eyes, with horses instead of the camels that would have been more historically accurate, in a sumptuous and ceremonial procession making its way to the walls of Jerusalem.

of the peninsula. Sometimes it followed the mountain route. Sometimes it followed what has become the modern route along the coast. Once past the the Gulf of Aqaba, it split again into lesser routes at the ancient city of Gaza. Some of these routes took the perfumes eastwards towards India. Others took them west towards the great Egyptian port of Alexandria. The Queen of Sheba's caravan, loaded with riches, followed a track northwards towards the capital of the greatest king of the region: Solomon.

Myrrh was, in the ancient world, a perfume, a gift to the gods, one of various incenses, a preservative and a cure for wounds and skin ulcers. It was also an aphrodisiac. The other commodities that loaded the Queen of Sheba's camels, things we now think of as kitchen spices, were also, for her, aphrodisiacs. The Queen's state, if it really was the ancient Saba and not on the other side of the gulf in modern Somalia (its location is disputed), was rich because of them. The region has an excellent climate. The lands then were well watered, thanks to the brilliant water engineering techniques brought from the city states of Mesopotamia in the distant past. The terraced hillsides were rich in many crops, but there were also vast groves of the thin and scrawny trees, whose barks, slit, gave rise to two of Saba's greatest commodities of the time: frankincense and myrrh. Myrrh (*Commiphora guidotii*) and frankincense (*Boswellia papyrifera*) appear in the Song of Songs (probably written long after both Solomon and the Queen of Sheba were dead), where in its heady stanzas they symbolize the bliss of sexual union: 'A bundle of Myrrh is my well-beloved unto me; he shall lie all night betwixt my breast'; and, later in the same poem: 'Until the day break, and the shadows flee away, I will get me to the mountain of myrrh and to the hill of frankincense.'

Perfumes are still erotic, still related to human sexuality and the pheromones of desire. They are still much used in the East, where men and women find that the sumptuous odours of olibdanum, myrrh, musk and rosewater set pulses racing. The West's idea of Eastern sensuousness, much envied in the West, particularly in the nineteenth century when chill air and Christianity had cooled ready access to sensuality, is often associated with the East's heady perfumes. In Europe, even the wonderful perfume of jasmine, lemon blossom and some lilies were thought to be bad for polite womenkind; perfumes were said to make them faint, or at least give them headaches. Polite men used the word 'spicy' to mean 'erotic', and even the phrase to 'ginger things up' could raise a smirk.

Many aphrodisiacs operate more in the mind than they do in the body. However, myrrh is more than a spicy and arousing smell. It, like ginger and cinnamon, is a mild stimulant too. Causing the pulse to increase and the brain to speed up, it might well – helped by a strong belief in what myrrh ought to do – have the desired effect. Perhaps the Queen of Sheba had this in mind. Solomon was reputed to have many hundreds of wives and concubines at his disposal. She may have needed to create an impact, and to have aphrodisiacs in plenty. Her intended mating with the sun – her spiritual life was centred around moon and sun god cults – may have worked. An Ethiopian epic, the *Kebra Nagast* (The Glory of Kings), which claims that the Queen of Sheba's lands were in Ethopian territory, also claims that she bore Solomon a son called Menelik, who on manhood returned to Solomon, who appointed him to rule all Africa.

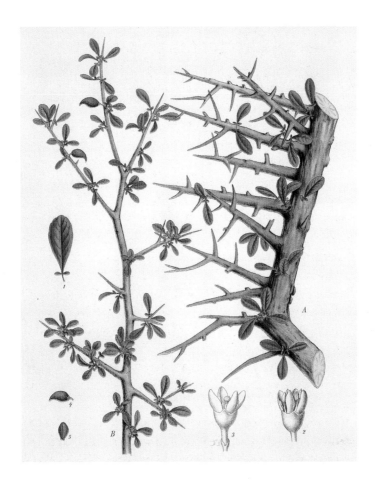

Myrrh (*Commiphora guidotii*) has spines and resin packed with fungicides and bactericides to deter grazing animals. It has been a valuable crop for many thousands of years: humans gash its stems and harvest the dried beads of resin, called 'tears', that ooze from the cuts.

Though frankincense and myrrh were probably the product of the Queen of Sheba's lands, the other commodities she took to Solomon came into her hands by trade. What had created her state's riches had been the domestication of the Arabian one-humped camel, or dromedary, which made trading journeys across deserts possible. Thus ginger, rubies and emeralds, which had arrived on small boats plying between the easy anchorages of Saba and the western seaboard of India, could be traded on northwards. Trade across the Indian subcontinent was already well established, so as well as India's own spices, gems and drugs, commodities could be brought from further east still: cinnamon from trees in the jungles of Ceylon, perhaps even nutmegs, mace and cloves from the unimaginably distant Molucca islands. The trade generated such wealth that great palaces and gardens were built along the southern and western coasts of the Arabian peninsula.

Wealth brought envy from more earthbound sources. There were early attempts at subverting the region's economy. Hatshepsut, Queen of Egypt in the fifteenth century BC, with the full titles and regalia of a pharaoh, sent an expedition to the now unknown Land of Punt, probably on the southern coast of Arabia, to search for incense and spice trees. Hatshepsut wanted living plants at Thebes as homage to the god Amon. She may also have wanted a cheaper source of supply. Cinnamon was important to her because it was a major ingredient in Egyptian perfumes, and a vital gift to the gods. Myrrh perhaps would also make a suitable gift. Her expedition returned to Thebes with thirty-one myrrh trees that were still alive. However, the Theban climate, so perfect for opium, was unsuitable for myrrh and it didn't become an Egyptian crop.

The Queen of Sheba's great caravan may also have had in its baggage something that lasted much longer in the aphrodisiacal stakes than ginger or cinnamon or even myrrh. It was certainly well known as an aphrodisiac by Roman times, and it was an ancient and important ingredient of love potions in Arabia Felix. This was the nutmeg, seed of a tree named *Myristica fragrans*. It is a native of the Spice Islands, now more correctly called the Moluccas, a group of volcanic cones, mountainous, fertile and very humid, that emerges from the seas of eastern Indonesia, between Sulawesi and New Guinea. When the nutmeg seeds ripen, they are partly surrounded by a translucent scarlet structure called an aril. This seems to play some role in making the seed more attractive to whatever animal in the Moluccan jungle acts as the plant's distributor. In the human world, it was realized that the aril contains different fragrances from the nut itself. Stripped away and dried, it becomes the spice called mace. This, too, has been used as an aphrodisiac.

In India, the Moluccan nutmeg had already been part of the medicinal and culinary flora for a thousand years before the Queen of Sheba's time. The Vedas describe it as warmth-producing, stimulating and good for digestion. It is also a hallucinogen, and this too was known – one of the synonyms for nutmeg in Ayurveda is 'mada shaunda' or 'narcotic fruit'. It was in widespread use as an aphrodisiac well into the nineteenth century, its use for that purpose having spread to Egypt, to Rome and into the Arab world, where it was used from the first century AD. Not all its uses were excitant. In traditional Indian folk and domestic medicine, nutmeg was and is used in small quantities to induce quiet in irritable children. In recent times, British nannies and mothers grated it into cups of warmed milk to help children sleep. Perhaps they still do. It is difficult to see how something narcotic and sedative, and even antispasmodic, could also stimulate sexual desire. As with many supposed aphrodisiacs, belief in its effects had to be powerful enough to overwhelm the plant's real effect.

Nutmeg's influence, both as spice and aphrodisiac, spread to medieval Europe. It gradually became a genuine folk remedy, and nearly every herbal contained a summary of nutmeg's virtues. By the eighteenth century, every gentleman carried his own nutmeg and silver nutmeg grater; or perhaps even then the bedroom was as much in mind as the trencher. More recently, nutmeg has become used as an abortifacient and as an hallucinogen when others are not easily available. Even though it is for sale in every supermarket, it can be a dangerous plant: eating the ground equivalent of two nutmegs can be fatal.

Queen Hatshepsut, seen in a granite relief. Astute and elegant, she explored the possibility of growing myrrh and other spice and incense plants at Thebes, rather than importing them from Arabia and India. Her plans failed; incense, and the kohl for her make-up, remained costly.

The two spices mace and nutmeg must have reached first the Mediterranean region and thence the rest of Europe using the Incense Road. That ancient route was exceptionally difficult, yet the drive for spices and stimulants was strong enough for traders to endure its many toll points and hazards. Traders worried away at the possibility of an alternative land-based journey, from the East to the trading cities of Turkey, where goods could be more easily protected against the wilfulness of nature, the greed of pirates and the customs dues of entrepôts. There were great obstacles: immense ranges of mountains – the Himalayas, the Pamirs, the great ranges of Tianshan, Karakorum, Kush. There were deserts such as the Gobi, a thousand miles across, and one so hostile that it was called the Land of Death, the Takla Makan. The Incense Road was gradually usurped in importance by another, a new overland route that eventually became hugely influential in the development of human civilization. In the nineteenth century, historians called this new route the Silk Road, even though its most important commodities were spices, medicinal plants and ideas. Bales of silk fabric were bulky, heavy and easily damaged and were certainly carried along it. Nevertheless, silk was not its most important commodity, however much it appealed to the imagination of later historians. In fact, the first Chinese silks began to appear in the West much earlier than the date usually thought of as marking the start of trade along the Silk Road – the usual date given for its formation is around 300 BC. Some rare Egyptian mummies were wrapped in silk around 1000 BC. Silk has been found in German graves dated to 700 BC, and in some Greek ones from the fifth century. European, Persian and Central Asian silk goods have been found in some Japanese tombs of similar dates. Perhaps a few merchants had created some early way of travelling a proto-Silk Road.

The formation of the Silk Road began at both ends. In the West, its growth eastwards from Anatolia was expanded by Alexander's forays towards India. This development was not in itself sufficient to drive the route through the mountains and terrible deserts that lay further east. For that was needed the immense wealth and military power of Rome, then rapidly developing. The Roman wars in Parthia from 53 BC established routes through Turkey and on into the easier terrain of Persia and Syria, providing adventurous merchants with a springboard to the further east. In the East the emperors of China's Han dynasty began to explore westwards. They found wonders: new varieties of horse, the Gandhara culture and its Buddhism, and exotic objects and substances from further west still. By 150 BC, ways of passage were created around the Takla Makan desert, to both north and south. By the time the Romans were fighting the Parthian King Mithridates in Turkey, the Parthians were beginning to see Chinese traders who had arrived in their lands from the East.

The greatest age of the Silk Road was in the eighth century, the time of the Tang dynasty in China. As the eastern starting point of the route, the old Qin capital, Changan, developed into one of the largest and most cosmopolitan cities of the time. By AD 742, the population had reached almost two million. The AD 754 census listed 5,000 foreigners: Turks, Persians, Indians, even Japanese, Koreans and Malays. Rare plants, medicines, spices and other goods from the West were to be found in the city's bazaars. The route, astonishing in its length and in the severity of parts of the journey, was entirely overland until the western coast of

Myristica fragrans (illustrated here by Elizabeth Blackwell) yields nutmeg and mace. Such spices drove the creation of trade routes through the most hostile terrains of central Asia.

OVERLEAF Spices from the tropic Far East first reached the West using either the sea route across the Indian ocean, landing in southern Arabia, or the Incense Trail up its eastern seaboard. This painting of Arabia and India is from the *Miller Atlas* by Pedro Reinel, *c.* 1519.

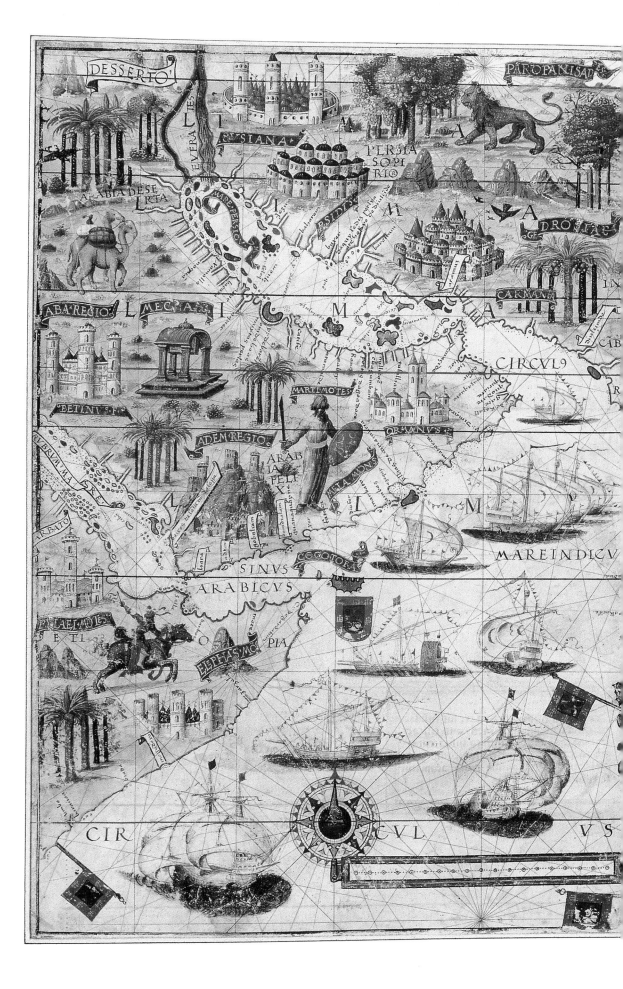

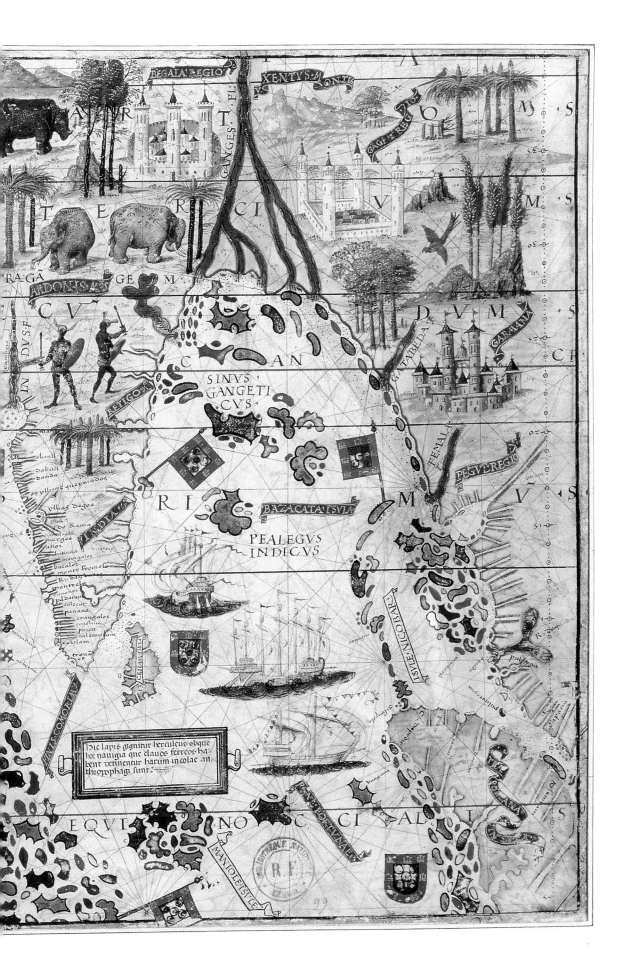

BEGALA REGIO

XENTVS MONTE

GANGES FL.

GAGE REGI

A R T CI V

TER CI

RA·GA GE M
ARDONIS M'S

CV

DVM

GARAIANA

S

CI

SINVS GANGETI CVS

CAN

GATABEDA

CAMBAIA

IN DVSF

RETIGO

FEMALA

PEGV REGIO

RI

BAZACATA ISVLA

M V S

PEALEGVS INDICVS

INDIA

ISVLE NICOBAR

CELLA·CASTA

Hic lapis gignitur herculeus obque hoc nauigia que clauos ferreos habent terinentur harum incolae anthropophagi sunt.

FINIS COMORINI

EQVI NO C CIAL I S

FINIS BORIZONALA

MANIOLE ISVLE

99

Turkey. In the overland sections, it generated wealthy entrepôts: as well as Changan, Samarkand, Caesarea (modern Kayseri) and a hundred more. Where it reached the Mediterranean, it gave rise to cities richer still: Ephesus, Pergamon, Smyrna to the west, Alexandretta to the south. Some of these great cities, where traders profited from medicinal plants from Western traditions going eastwards and Eastern ones arriving in the West, became centres of medical theory – one of these being at Pergamon, where developed the huge medical school, the Aesculapium, where Galen studied, and which had one of the greatest of all ancient medical libraries.

The library remained great until Mark Antony gave it to Cleopatra as a wedding gift to restock the sacked library of Alexandria, after its first burning. The Romans back in Rome who disapproved of Mark Antony's heroic erotic obsession thought Cleopatra an enchantress. Whether or not she was, it is likely that two plants were used in her lovemaking with Antony. One of these, the nutmeg, would have reached her along the Silk Road. The other had long grown in her own kingdom of Egypt. This was the mandrake.

In general, like fashionable medicines, most aphrodisiacs were more highly thought of if they came from distant lands. After Columbus reached America, what are now ordinary vegetables were then strange enough to give the consumer the right sort of sexy frisson. Aubergines, sometimes phallic in appearance, were found arousing. The now humble tomato, too, was a danger to the chaste, and called the love apple. Though it became an essential part of cooking in the Mediterranean region within a century or two of its introduction, it was frowned upon in northern Europe. In Presbyterian Scotland it was hardly consumed until well into the twentieth century. Both plants are related to henbane and belladonna, and related to the plant most often accused of being Cleopatra's bait.

The mandrake was, and is, native to much of Europe and the eastern Mediterranean. There are several species, the most important of which is *Mandragora officinarum*. In the garden it is a rather modest plant, with a wide rosette of spear-shaped leaves that sprout from the rootstock every spring. As the leaves expand, so do the long flower stalks, which each bear a pale grey-green flower. If fertilized, these go on to produce a glossy green fruit the size of a hen's egg. The flesh is sweet and yellow, and not especially poisonous. The rest of the plant contains the usual family of alkaloids: scopolamine, atropine and so on, plus another, now called mandragorine.

The plant has been used medicinally since ancient times – indeed the common name may derive from a Sumerian phrase for the 'plague god plant', and suggests it was once used as a hoped-for cure for early urban diseases. The Sumerians also used it for gut and toothache, and to ease the pains of childbirth. It also seems to have been associated with a rich mix of magic,

A plaque from the lid of a coffer showing Tutankhamun (*c.* 1370–52 BC) and his wife standing in a garden amongst fabulous papyrus plants. The frieze beneath them shows a mandrake plant in full fruit. Mandrake had a vast range of uses, but here it is probably associated with sexual arousal and fertility. Much later, Arabic doctors called it Satan's apple.

sex, madness and death. That the ancient Greeks believed it was an aphrodisiac is suggested by the fact that their goddess Aphrodite claimed it as hers, and her adherents made use of its various properties. The enchantress Circe, who so nearly ensnared Odysseus and turned his men into swine, was supposed to make, or at least flavour, a wine of mandrake called Circaeon. Odysseus, though, was immune to her charms and her wine, for he was protected by a herb given to him by Hermes. The mandrake appears in the Bible, in a confusing passage in Genesis where the elder wife of Jacob buys a night with her husband from his younger wife. The bargain is made for some mandrake plants, which were also believed to give fertility. In ancient Egypt mandrake was widely used and grown, and is shown on many wall paintings. As well as a painkiller, it was an important aphrodisiac. Its bright green fruits, taut and glossy, had a symbolic erotic significance throughout pharaonic times.

The Greeks had some good stories about mandrake. Theophrastus wrote: 'It is said that one should draw three circles round mandrake with a sword and cut it with one's face towards the west; and at the cutting of the second piece one should dance round the plant and say as many things as possible about the mysteries of love … The leaf of this mandrake, used with meal, is useful for wounds … and for love potions.' He goes on to describe dried beads of its root being used for amulets. Roman usage was similar, and its anaesthetic properties became important. A person about to undergo surgery was given a piece of root to chew. Apuleius, a second-century Roman writer, writes: 'For … devil sickness or demoniacal possession, take from the body of the said … mandrake by the weight of three pennies, administer to drink in warm water as he may find most convenient. Soon he will be healed.' The association with demons is long-lived. Arab doctors called it Satan's apple. In Europe, the uninitiated were warned against harming the mandrake's root. Anyone who pulled one out of the ground, it was believed, would stir the fury of the demon who inhabited it: the demon's shrieks of anger would be so horrible that the transgressor would die on the spot.

The root of the average mandrake plant looks like a pale carrot. Sometimes, also like a carrot, it forks. When forked, it was, in the past, supposed to resemble the human body. It was believed that the relevant piece of the root could cure the analogous piece of the human. Some roots were supposed to be male mandrakes; more rarely were the roots female. Herbals commonly show the mandrake as a rather gnarled male with a long beard. The 'female' mandrake usually has a surprisingly bushy head of hair. The drakes, dried, were valuable objects of magic. They were often faked. Bryony (*Brionia dioica*) roots, for instance, were cut into fancy shapes and passed off as mandrake. Clever fakers grew their bryony plants in special moulds, so that the expanding root took on the desired forms. The mandrake mannikins were believed to attract good luck, good fortune and perhaps good sex. They were carried secretly, as the owners feared being accused of practising witchcraft. To make a mannikin more realistic still, grains of millet were embedded into the face as eyes. Italian ladies were reputed to pay as much as thirty golden ducats for these 'puppettes' or 'mammettes'. The English herbalist Turner (*c*. 1508–68) alludes to these dolls, saying 'they are so trymmed of crafty theves to mocke the poore people withall and to rob them both of theyr wit and theyr money.' He adds:

'Of the apples of mandrake, if a man smell of them thei will make hym slepe and also if they be eaten. But they that smell to muche of the apples become dum … thys herbe diverse wayes taken is very jepardus for a man and may kill hym if he eat it or drynk it out of measure and have no remedy from it … If mandragora be taken out of measure, by and by slepe ensueth and a great lousing of the streyngthe with a forgetfulness.'

By the time William Shakespeare was writing, his audience was aware enough of mandrake for him to use it as an ill-fated sleeping potion in *Romeo and Juliet*, which was probably written in 1597. At least some of the audience would have believed that they could dig up a plant only by tying a fairly large dog to the root: the dog was encouraged to make a run for it, thereby hauling out the root too, and in the ensuing uproar, the dog died and its master harvested the root. However, by the same date, rationalists were beginning to disbelieve the legends about mandrake. Gerard, with no time for astrology or magic, pours scorn: 'There have been,' he writes, 'many ridiculous tales brought up of this plant, whether of old wives or runnegate surgeons or phisick mongers, I know not, all which dreames and old wives tales you shall from henceforth cast out your bookes of memorie.' Splendid. In the next century, Culpeper explains that the mandrake is governed by Mercury. More importantly, he points out that 'the root formerly was supposed to have the human form, but it merely resembles a carrot or parsnip'.

However, the root is still used as a lucky charm in some southern European and African cultures. Its properties are also being explored by seekers of hallucination. It needs to be treated with extreme caution. At too high a dose, the individual can experience violent vomiting and diarrhoea, lasting anything up to ten hours. There is an occasional death from overuse. In America, even the terminology is dangerous. The 'American mandrake' can be a different plant altogether, the handsome *Podophyllum peltatum*, which has fleshy roots and glittering green fruits. This plant is a very powerful cathartic, and needs even more careful handling than the true mandrake.

Chinese doctors had a large range of aphrodisiacs, love philtres and other aids. The manual, or compilation, produced by Tung Hsuan Tsu some time between the fifth and seventh centuries AD proclaimed that 'Those who understand the nature of sex will nurture their vigour and prolong their life. Those who treat its principle with contempt will injure their spirit and shorten their life.' For those whose vigour quailed at such slogans, the list of plants was wide. The Chinese foxglove (*Rehmannia elata*), used in many medicinal recipes, was an important species. So was *Schisandra sinensis* and the strange little dodder *Cuscuta japonica*. The list also included various exceptionally poisonous aconites, and a relative of the European milkwort, *Polygala japonica*. All have been used for at least two thousand years, though there is no modern research on their effects. However, one plant was used beyond all others, and has been extensively researched: *Ephedra sinica*, or ma-huang, which is a singular stimulant, containing ephedrine, a substance close to amphetamine.

A plant of the dry steppes, ephedra is first mentioned in the classic Chinese herbal of the Divine Ploughman Emperor, Shen-Nung, of the first century AD. It is included in the list of 365 herbs that still comprise the basis of the modern Chinese materia medica.

While it was also supposed to cure everything from typhoid fever to swelling of the ankles, in particular it provided and still provides energy and increased alertness. These are the effects of ephedrine, which are exactly the same as our own self-produced hormone, adrenalin. Some people find ephedra arousing, particularly if that is the effect they are expecting, and it was used as an aphrodisiac by the Mongolian tribes. The genus is also found in American dry lands, and native Americans found the plant stimulating. The plants from the deserts of the south-west, are often called, humorously, Mormon or Brigham tea after the founder of the Mormon community who had more than twenty wives. In present-day America, extracts of *Ephedra* species have become widely used as a general 'pep' pill, to improve sexual as well as non-sexual athletic performance, and as an aid to slimming. Many find that it makes them feel too nervous and restless to concentrate on making love. It may be so widely used in Chinese medicine simply because it makes the patient feel livelier. The alkaloids ephedra contains are very powerful. Common side effects, at high doses, are unpleasant: intense restlessness, anxiety, dizziness, insomnia, tremor, rapid pulse, sweating, respiratory difficulties, confusion, hallucinations, delirium and (very infrequently) convulsions. Individual responses to ephedrine are also very variable, and there have been a number of deaths attributable to the plant's use. It is especially risky for anyone with high blood pressure or a weak heart. The plant drug is becoming widely regulated, one of the reasons for this being that its alkaloids are used in the production of illicit methamphetamine, or 'speed'.

Rehmannia elata, illustrated here by Henry Moon, is called the Chinese foxglove, though it is not related to the common foxglove (*Digitalis* spp.). It was once much used as an aphrodisiac. An ingredient in many Chinese medicinal recipes, it is believed to nourish the yin, and is used in a variety of conditions including stomach cancer, broken bones, tinnitus, wet dreams and something called steaming bone syndrome.

In the quest for Flora, goddess of flowers, to encourage ourselves and our lovers to greater efforts, we have ransacked jungles and deserts for new discoveries. Some 'finds', though, are ancient ones revamped, or new ones quickly claimed as miraculous. One such find, receiving much attention today, comes from a tree of the tropic rainforests of Nigeria, Cameroon and the Congo. *Pausinystalia johimbe* contains an alkaloid called yohimbine. It is used by local shamans to induce visions. At lesser doses, the general population use it as an aphrodisiac. Over the last century, it has been enthusiastically taken up all over the globe in that role, especially by ageing men worried about their reduced drive. It is now widely used in so-called natural mimics of Viagra. The demand for it is huge, and hundreds of tons of bark are shipped every year to America and Europe. It is all collected from the wild, but there are only about two trees per acre. The species is already almost extinct, yet the wild bark costs less than 50 cents a pound. Some plantations have been established, but it will be years before they become productive.

Like many stimulants, yohimbine increases adrenalin and norepinephrine in the bloodstream. This effect can radically increase blood pressure and heart rate, but in some people, and at low dosages, it appears that yohimbine can lower blood pressure, probably by dilating blood vessels and so reducing vascular resistance. As this takes place particularly in the lower abdomen, its occasional effect on the male organ may be explained. It may cause some stimulation in women, although in one trial only a third of subjects reported beneficial effects. It can also induce a long list of serious side effects, and block or enhance the effects of other drugs being taken. In any case, the chances of finding yohimbine are slim. Perhaps because the wild tree is getting rare, analyses of commercial yohimbine extracts show great variability in content as it becomes more and more frequently adulterated. The vast majority are largely devoid of any effective level of yohimbine at all. Concentrations typically range from zero to almost 500 parts per million. The bark contains about 7,000 p.p.m. Other plants contain yohimbine: one is an American periwinkle, *Catharanthus lancei*, but the 'down' after the stimulus is reputed to be nasty.

Some plants used as aphrodisiacs have not yet even been examined by chemists – the lovely scrambling cactus sometimes called 'Queen of the Night' (*Selenicereus grandiflorus*), for instance, native to Mexico, the American south-west and Jamaica. The huge trumpet flowers, creamy white, jagged petalled and often more than a foot across, have a bewitching smell, and last just a few hours. Half a gram of flowers is supposed to be good, and even better if taken as an alcohol extraction, using whisky, brandy or rum.

The most surprising plant of all is one that produces substances that are remarkably close to the human sex hormones testosterone and progesterone. It is not clear if these cancel each other out in the human body when the plant extract is taken; some herbalists fudge and say that it 'balances the hormones'. The Amerindians of Central America used the plant, at least by the time that European settlers arrived, to promote male behaviour and male potency. The plant is sarsaparilla (*Smilax officinalis*, and other species). In the 1930s, several pharmaceutical companies became interested in stories that Mexican Indians took a root extract to increase their machismo. The companies' chemists discovered that sarsaparilla contained impressive amounts of testosterone. The plant became immediately important because the only previous source of testosterone was that extracted from bulls' testicles obtained from slaughterhouses. Supplies were therefore extremely expensive. For a while sarsaparilla became the main source of the substance worldwide. Wild smilax soon became scarce, and the hormone became expensive again. However, before it became a farmed crop, chemists discovered that they could synthesize testosterone from cholesterol.

Away from the demands of machismo, sarsaparilla has been claimed to have many other dramatic effects. It has been used to expel gas from the intestines, and to increase perspiration and the volume of urine produced by the kidneys. Having a reasonably pleasant taste, it has also been used to flavour root beer and various commercial soft drinks, though it is not clear if testosterone survives the brewing and bottling process. If it does, the other hormone may do so too. It has also been used as a hair restorer by balding men, as well as an alleviant of rheumatic pain.

More mysteriously, sarsaparilla has been claimed to be a general antidote for poisons, especially mercury. Though now rare, mercury poisoning was common in the days when mercury was used as a cure for syphilis. However, sarsaparilla was used by native Americans not only to boost their sexuality but also, as they knew nothing of mercury, to cure the syphilis that was sometimes the cost of sexual pleasure. News of that use gradually reached the desperate West. In late sixteenth-century London, Gerard knew of the plant, but didn't yet know of its supposed antisyphilitic property. By the eighteenth century, it was available in London shops, called 'zarza parilla', and was recommended by the College of Physicians. It was even believed to cure infants who had been infected with syphilis 'by their nurses', though presumably nurses often took the blame for infections in fact passed on by mothers. The information must have followed the journey of the disease to China. There, it is still highly rated as a cure. The genus *Smilax* is well represented in China, and Chinese specialists use local species.

The plant called American sarsaparilla is unrelated, though the real sarsaparilla is equally American. The 'fake' plant is *Aralia nudicaulis*, distantly related to the ivy, but not, like *Smilax*, to the lily. Also known as false sarsaparilla, shot bush, small spikenard, wild liquorice and rabbit root, it was used by the Cree Indians as a cure for syphilis and as an application to recent wounds. It is much more poisonous than smilax.

In the West, the disease of syphilis has not always been the dark reward of the pursuit of spicy lives. Though there is some dissent, it is extremely probable that syphilis did not exist in the Old World before 15 March 1493. On that day, Christopher Columbus landed once more in Europe, after an exceptionally momentous expedition. Much miscalculating the size of our globe, he had been hoping that, setting out westwards, he would eventually land somewhere near India, China or, most desirably, the Spice Islands. His backers hoped for a route that would be faster, and perhaps less hazardous, than the newly discovered one that rounded the southern tip of Africa. Having set off into the Atlantic, his landfall was not, of course, the Indies but an entirely unknown and largely unsuspected continent. Though he asked for, and eventually got, gold from its natives, he didn't find spices. He did find many other plants that figure in this book, as well as ones that went on to be vastly important food crops. He also found, it seems, a terrible disease. When he and his crew arrived back in Spain, they had been at sea for two months. The syphilitic chancres that would have appeared on any of his crew who had accepted the kind hospitality of the newly discovered 'Indians' would have vanished by then and the sufferers would have thought they were now quite well. In fact they were bearing the new scourge.

Syphilis is caused by a tiny blood parasite, a spirochaete called *Treponema pallidum*. In the West there were already similar diseases caused by closely related species of *Treponema*. Bejel, yaws and pinta were, and are, scourges transmitted by direct skin-to-skin contact, or the sharing of clothes, cups and so on. They are not specifically related to sexual encounters. Though unpleasant, they are less terrible than the syphilis produced by *Treponema pallidum*. The other treponemal diseases attack fingers, toes and sometimes internal organs. None goes on to attack the brain. Syphilis was certainly present in the New World long before Columbus arrived, and may have evolved from yaws, perhaps 1,600 years ago. The first unambiguous descriptions of syphilis in the Old World appear in France around 1500, seven years after Columbus's return. Though these descriptions may just reflect growing medical knowledge and the ability to differentiate syphilis from other similar diseases, recent researches suggest they really do signal its recent arrival from the New World.

In Europe, the disease created terror. It spread fast, becoming an epidemic by the beginning of the sixteenth century. As it spread it was variously called the French disease, the Venetian pox, the Neapolitan disease and so on. Muscovites called it the Polish sickness. The Poles called it the German disease. Everyone blamed someone else near by. No one seems to have suspected the Americas. Travellers were blamed. Prostitutes were blamed. Soldiers were blamed. Certainly, the mercenaries employed by Charles VIII during his Italian campaign in 1495 seem to have become infected in some quantity. Those who survived the fighting returned home relieved to be still alive. They had a wild time and so spread it quickly.

The first city to think the disease was a new plague was Lyons. The ill were forced out beyond the city walls in March 1496. It is hard now to know what the criteria for exclusion were, for the disease develops slowly and the first signs can, especially in women, be invisible. Ten years after the Lyons scare, by which time the disease was present throughout Europe, some sufferers were progressing to the next phase of the disease, and dying terrible deaths, insane. The horror intensified when it was noticed that babies were being born with the disease, already chancrous and often with their faces badly deformed.

Doctors and apothecaries tried to help, employing the usual ferocious armoury of purgatives, emetics, bleeding and cupping. Plants were used in complex formulae. The rare and the exotic were, as usual, preferred to locally available materials. The lovely blue-flowered American lobelia, now called *Lobelia siphilitica*, was first imported to Europe as a cure. Gerard suggested guaiacum, the gum from a Jamaican tree (*Guaiacum officinale*, or 'lignum vitae'), as the most ancient and powerful antidote. Another common name for the plant was Indian pox wood: it was one of the plants found by Columbus, and long used in the Caribbean. Gerard also describes and illustrates the 'China root', which he says was good for the long-standing

Andreas Vesalius, a Belgian educated at Louvain and Padua, became the most important Renaissance anatomist, influencing future developments with seven volumes on the structure of the human body, illustrated with engravings taken from his own drawings. This painting of 1564 is by Jan von Calcar.

disease and more powerful than either 'guiacum' or 'sarsaparilla'. However, he didn't realize that 'China root' was in fact a Chinese species of sarsaparilla. China root was widely imported. In 1546, the anatomist and doctor Andreas Vesalius wrote an *Epistola* on the discovery and therapeutic use of China root in the treatment of syphilis. The plant does have indeed some anti-inflammatory action, so it may have helped the patient. All these drugs were expensive. Cheaper and more homely, Gerard's other cures for 'French pox' included a species of dodder (probably *Cuscuta epithymum*) that grew only on thyme, water germander (*Teucrium scordium*), the distilled flowers of *Viola tricolor*, red star thistle (*Centaurea calcitrapa*), the yellow fumitory (*Corydalis lutea*), myrobalan water and soapwort (*Saponaria officinalis*). He quoted one apothecary as saying that the taste of soapwort was so horrible – it is – that it could only be used for poor people. None of these plants would have attacked the treponema.

The disease soon spread to the East. In China, it appeared first in the southern region of Guangdong in the early sixteenth century. Sufferers were isolated, as if they had leprosy. Chinese doctors used *Smilax china*, and indeed still advocate the use of China root. In India, cannabis was used for all venereal diseases. It perhaps helped with the pains of the secondary and tertiary stages of syphilis. In eighteenth-century Japan, the disease was rife. The fact that European doctors were allowed far more access to the country and its plants than European merchants allowed Carl Peter Thunberg (1743–1828), a favourite pupil of Linnaeus and a doctor specializing in syphilis, to make the first big collection of Japanese plants. However, he abandoned the use of medicinal plants in the disease's treatment, and had great success using mercury compounds to treat Japanese sufferers, who were hugely thankful.

One of the reasons that so many plants were used in attempted cures for syphilis, and indeed why mercury or arsenic quite often appeared so effective, was due to the nature of the disease itself. Syphilis has four stages. The first signs appear a few weeks after infection. The specific area of infection develops a chancre or sore. Even untreated, this sore vanishes within a few months. At this point, the hopeful doctor or apothecary can claim a cure. The next stage begins six to eight weeks after the chancre's disappearance. The sufferer feels tired and headachy, and may have swollen lymph nodes in armpit and groin and a sore throat; a rash may develop. These symptoms, as treponema now affects the entire body, can last from three to six months, vanishing and reappearing apparently at random. Then all symptoms vanish. This is phase three. Doctors may once again believe that they have achieved a cure. This is the 'latency' stage. It can last for one year, sometimes a whole lifetime. More than half of all sufferers live the rest of their lives without the disease progressing to the infamous final phase. Nor are they particularly infectious. With these patients, herbs or antimony, or arsenic or mercury, or whatever is used to treat it, may seem to have triumphed. Once again, the hopeful doctor or apothecary can claim a cure. For sufferers in whom the disease progresses, however, treponema gradually infests the bone marrow, the lymph glands, vital organs and the central nervous system. Ugly and painful lesions develop all over the skin, bones and vital organs. Bones may get eaten away. The spirochaetes may concentrate on the arteries and valves of the heart, destroying both. They may attack the brain. Horrible personality changes can occur and the end result can be a helpless maniac suffering from what was called the 'general paralysis of the insane'. Even today, doctors can be misled into thinking that some of these strange expressions of the disease are due to some other cause. The great Canadian doctor Sir William Osler (1849–1919) wrote: 'He who knows syphilis, knows medicine.'

Mercury was the earliest moderately successful chemical treatment for syphilis. Cinnabar, an oxide of mercury, had been used in the treatment of leprosy and various superficial ulcers from the 1300s. Its use on syphilitic chancres was an obvious progression. Giorgio Sommariva of Verona seems to have been the first physician to use it in this way in 1496. Jacopo Berengario da Carpi then became famous in Italy for using pure mercury, not its oxides, which he managed to do without killing his patients. The metal was pulverized in ointments or in liquids to be swallowed. Mercury vapour baths were also widely used. The effects of the poison on the human were devastating, but it sometimes completely killed the spirochaete. Mercury continued in use for the next three hundred years in increasingly liberal amounts. By the 1800s, mercury was being used on nearly any ulcer found, and many patients were more injured by the treatment than the ailment. Mercury was not effective on late-stage syphilis and its deep-seated lesions. By the 1840s, potassium iodide had come into use. It turned out to be amazingly effective, even on patients in the final stages of the illness. Some North American doctors tried to achieve the best of both worlds, by combining the potassium iodide with the native American remedy sarsaparilla.

In 1905, the syphilis organism was discovered. By 1908 Paul Ehrlich was experimenting on syphilitic rats, using a huge variety of different arsenic compounds. At number 606, he found one that largely destroyed the syphilis without also destroying the rat. In 1910, the substance, called salvarsan, was introduced to the world. A great fear, and a terrible disease, had met its match. Moralists were outraged, believing that a cure would encourage sexual promiscuity – echoing the moralists of the Renaissance who saw the disease as a punishment for chasing Venus. Nowadays, penicillin has superseded salvarsan. Yet *Treponema pallidum* is evolving and resistant strains continually occur.

Though syphilis was, like today's HIV, a major and terrible fear among the sexually active, there were also lesser worries and irritants. Gonorrhoea was one of these. This was an ancient Western disease, often thought of not as a disease at all but as an 'involuntary flowing away of the seed'. It was commonly called 'running of the reins', for the unpleasant and painful exudate was thought to be merely excess semen. Seventeenth-century cures ran from straight knotgrass (*Persicaria* spp.), used to soothe the inflamed 'secret parts of man or woman', to some rather pleasing recipes. One ran: 'The roots of Comfrey in number four, Knotgrasse and the leaves of Clarie, of each a handful, being stamped all together, and strained, and a quart of Muscadell put thereto, the yolke of three egges, and the powder of three Nutmegs, drunk first and last is a most excellent medicine against a Gonorrhoea or running of the reins.' Patients must have got both drunk and high, though they would not have been cured. The roots of the yellow waterlily, turpentine resin, the pulp of rotten apples and bitter tamarind juice were all called into service. In eleventh-century China, the long-suffering gingko was tried, its nuts being eaten cooked. As with so many other plants, the nuts of gingko possess bactericidal compounds, here ginkgolic acids and ginnol. In South America, the infection was also widespread. In 1570, Gabriel Soares de Souza, a European observer, noted the Brazilian Guarani Indians using a plant to treat mouth ulcers. It was the jaborandi, once a common sub-shrub about three feet high, with smooth grey bark and large leathery leaves. In the 1630s two Dutch West India Company scientists documented Brazilian Indians using it as a tonic or panacea for colds and flu, and as a remedy against gonorrhoea.

There are no plants that cure love's diseases. There are no plants that are clearly aphrodisiac. But we have nevertheless wanted even more; we have hoped that plants might give us something magical – a love philtre that will bring to our beds, voluntarily, the silken cheek or the tawny beard we so desire. Every culture has claimed to find such plants, and in spite of every failure, nothing has diminished the fantasies of power and control dreamt of by the baffled would-be lover. On New Year's Day 1605, at the King's Revels, the newly crowned James I of England watched a play written by William Shakespeare ten years earlier. It was *A Midsummer Night's Dream*, in which Shakespeare refers to that dream plant:

> OBERON: Fetch me that flower; the herb I showed thee once:
> The juice of it on sleeping eyelids laid
> Will make or man or woman madly dote
> Upon the next live creature that it sees.
> Fetch me this herb: and be thou here again
> Ere the leviathan can swim a league.

> PUCK: I'll put a girdle around the earth
> In forty minutes.

Members of the audience would have smiled, for all would have known that to put a girdle around the earth took three years. Less than half a century before, the dream of unlimited access to erotic conquest, as well as a more savoury kitchen, had fuelled an incredible journey in search of the Spice Islands using Columbus's route across the Atlantic. Ferdinand Magellan believed that there must be a passage around or through the new continent discovered by Columbus. Like Columbus, he had miscalculated the size of the globe, and believed that the Spice Islands must lie only a few days' sailing from the new continent's western shore.

Magellan's five galleons set off from Spain in September 1519: the *Trinidad*, the *Santiago*, the *San Antonio*, the *Concepción* and the *Victoria*. The Spanish captains of four of them were already plotting to kill the Portugese-blooded Magellan. Mutiny, fires, murders, terrifying storms, disease and hostile natives all dogged the expedition. Nearly three years later, on 8 September 1522, a tattered *Victoria*, dulled pennants cracking in the breeze, tacked up the Guadalquivir River, to the quays of Seville. On board were only eighteen of the original 250 crewmen. Some had been executed for mutiny. Many had drowned. Many had died of disease and starvation. Some had been abandoned on the shores of Patagonia. Some had been captured and killed by rival Portuguese traders who reached the Spice Islands by sailing around Africa. Magellan himself had been killed in a battle between rival warlords on the Philippine island of Mactan in April 1521. The other ships were wrecked, burned, or captured. Amongst the eighteen survivors were the captain, Sebastian del Cana, and an Italian gentleman-explorer from Vicenza called Antonio Pigafetta. Pigafetta eventually published the story of the voyage as *Relazione del Primo Viaggio Intorno al Mondo*. The earliest printed edition, apparently a summary of Pigafetta's Italian manuscript, was issued in French by Simon de Colines of Paris in about 1525. The earliest Italian edition is of 1534.

As well as the few survivors, the *Victoria* contained huge quantities of spices: twenty-six tons of nutmegs, mace, cinnamon, cloves and sandalwood. Their value was so immense that they more than covered the cost of the entire expedition. They even paid for Sebastian del Cana's pension. His new coat of arms displayed two cinnamon sticks, three nutmegs and twelve cloves.

The sea journey to the Spice Islands, later accomplished more easily by other adventurers, eventually killed off the trade along the ancient Silk Road. Climatic changes were already driving sands over some of the most important trading posts along the road, and the dwindling trade made it no longer economic to risk the desert. Medicinal plants, doctors and apothecaries ceased travelling along it, and the easy linkage between Eastern and Western medicine came to a halt. In the West, a huge and chaotic dynamism, now called the Renaissance, was afoot which could not feed change into the stable and immemorial systems of China and India. But none of these immense events, the discoveries of new continents and new oceans, the developments of modern anatomy or the finding of new plants, diminished human desire to smooth the path of love.

c.

b.

a.

THE KILLING
PLANTS

A LINER steams slowly upriver. It is evening. On one of the decks, two men lean over the railing, eager for the first sight of their new city. One is middle-aged, ill-favoured, spectacled, anxious. The other looks much younger, fresh-skinned, eager. Other people on the deck walk swiftly by, wanting nothing to do with them. They have been seen holding hands. In a few hours, the SS *Montrose* slips in through a barrier of mist, and continues up the St Lawrence River, towards Quebec. A pilot cutter appears from the shadows, and comes alongside. A group of men board the larger boat. On the quayside further upriver, a car waits. The two voyagers who have been holding hands are soon arrested. One is Dr H.H. Crippen. The other is his mistress, Ethel le Neve. His wife, or what remains of her, lies beneath the floorboards of Dr Crippen's London house, 39 Hilldrop Crescent.

Hawley Harvey Crippen hadn't had much luck. He was born in Michigan, in the United States, in 1862. He first came to England when he was twenty-one, hoping to study as a doctor. He didn't manage this, though he returned with some sort of diploma in ophthalmology, endorsed by the Faculty of the Medical College of Philadelphia. Back in America, he studied more at the Ophthalmic Hospital in New York. He wandered about the US, practising in a number of larger cities. In Utah, during 1890 or 1891, his first wife died, and he sent his young son to live with his mother-in-law in California. When he returned to Britain, he was still unable to practise as a doctor, so he found employment as a dentist, and as a 'rep' for a quack medicine company.

The position was good, but Crippen, a strange and alienated man, had contracted a disastrous new marriage. Kunigunde Mackamotzki, or Cora or Belle Elmore, was a large and domineering woman. Believing herself to be an 'artiste', she wanted to sing Brünnhilde, but found it hard to get work even in the music halls. Wilful, extravagant and greedy, she constantly needed money. She got most of it from the lovers she flaunted at her husband. Crippen, constantly abused by his wife, fell in love with his secretary. The secretary, Ethel le

Hemlock (*Conium maculatum*), a near but lethal relative of parsnip and carrot, is found in damp areas throughout Europe and North Africa. Used for judicial killing by the Greek states, it has few positive uses, though that has not stopped its introduction to the Americas and New Zealand.

Neve, seems to have loved him in return. By December 1909, Belle was tired of him, and knew that he was in love with Ethel. She threatened to leave. Worse, she threatened to take their joint savings with her. On 15 December, she notified the bank of a forthcoming withdrawal. On 17 January, Crippen ordered five grains of hyoscine hydrobromide from Lewis and Burrow's shop in New Oxford Street. It was such a large order that the shop had to place a special request with their wholesalers. It was something they remembered. Crippen collected the order on 19 January 1910. He wasn't much good as a businessman or even a husband, and he was worse as a criminal. He signed the poisons book at the shop in his own name. He asked his office boy to buy several sets of clothes suitable for a young man. The letters to Belle's friends and admirer, purportedly from her, explaining her silence or her absence, were clumsily faked. The police had all the evidence they needed to charge him with murder.

The ultimate power over another human, or any other organism, is that of life or death. Many plants have been used to secure the latter. Some have been Janus plants. Others have been truly plants of darkness alone. They may fulfil the sentimental herbalist's fantasy of 'no side effects', but only because the taker lies dead.

Crippen had poisoned his wife with extract of henbane. The use of henbane, or its derivatives, as a poison stretched back to ancient Rome and beyond. It was one of the plant-based drugs used by the Greeks and Romans when going a-poisoning. Of these, the 'top ten' poisons were opium, mandragora, henbane, belladonna, thorn apple, hemlock, aconite, hemp, various poisonous mushrooms and yew. Yew was special. Death comes from

ABOVE Dr Hawley Crippen, seen here in a photograph of 1910, tried to escape from an appalling marriage by poisoning his wife Cora with one of the alkaloids from henbane (*Hyoscyamus niger*). He was caught fleeing to America with his secretary-lover and his trial became a sensation.

RIGHT In 399 BC, the seventy-year-old rationalist philosopher Socrates was tried by the Athenian court on the charge of encouraging disbelief in the gods, convicted by a pious jury and sentenced to death by hemlock. He is shown here in a painting by Luca Giordano (1634–1705).

respiratory paralysis. Pliny claimed that even sleeping or resting under the tree could be fatal, but the yew became harmless once a copper nail had been driven into it. Though these poisons were used for private purposes, they all had a public function too, being used for judicial killings and necessary suicides. When, around 399 BC, the philosopher Socrates was condemned to death and allowed to administer his own sentence, he took a cup of hemlock.

Hemlock (*Conium maculatum*) is a plant of hedgerows and ditches, growing all over Europe. It and its relative the water hemlock (*Cicuta maculata*) are exceedingly poisonous. Neither has been discovered to do much else other than kill. *Conium maculatum* has had occasional devotees amongst doctors, but, perhaps from tradition as much as anything, it has never been at all widely taken up. Socrates may, like others, have combined it with opium to make the fatal draught. Certainly, opium would have made his end pleasanter, as death by hemlock is not easy, for it causes a gradual paralysis of the body. Socrates would have noticed the loss of feeling in hands and feet, then in his limbs. Eventually, his rib cage would have become paralysed, and he would have died of asphyxiation. Hemlock has no effect on the brain, and without opium he would have remained entirely aware of what was happening until the end. Hemlock remained in use into the days of the Roman empire, a suitable dose being delivered to those whom the emperor decided he could do without. When Seneca was so marked, the hemlock failed to do its job. Either Seneca was too strong, or the dose too weak. He was smothered in a steam bath instead.

The alkaloids in hemlock are structurally similar to nicotine. As they are a powerful depressant, some doctors have tried to use hemlock as an antidote to strychnine poisoning. Hemlock was sometimes suggested in post-Roman medicine, combined with betony and fennel seed, as a cure for the bite of a mad dog, but while it may have reduced the intense spasms of hydrophobia, it probably would not have saved the patient. Greek and Arabian physicians used it for scrofula and cancerous tumours. The plant became widely naturalized in North America, which suggests that it must have had some use amongst settlers. It was only marginally taken up by nineteenth-century American doctors, who suggested it for calming children producing new teeth, various cramps and the psychological difficulties resulting in acute mania. In general, doctors there preferred other poisons to try, diluted, as cures, writing long-winded treatises on their marvellous effects and in doing so following in a long tradition of doctors who became enthralled by intensely poisonous plants.

The first Western book written on poisonous plants was by the Greek botanist, pharmacologist and doctor Nicander of Colophon (*fl.* 197–130 BC). He was born in Clarus, near the Ionian city of Colophon, and became a priest of Apollo. Much of his writings has vanished, but *Alexipharmaca* and *Theriaca* survive. The first describes venomous beasts, the effects of their poisons and suitable cures. It also includes 125 plant species. The second describes eleven plant poisons and their cures. *Theriaca* was an immensely influential work, and the word theriac eventually came to be used to describe an antidote to any sort of poison.

Theriac had exactly the same meaning as mithridatum, a substance of variable contents which, even though completely ineffective in all its various recipes, was in use for almost two thousand years. Mithridatum was named after Mithridates VI, Eupator Dionysius, sometimes called Mithridates the Great. He was a king of Pontus, part of north-eastern Turkey that had become autonomous after the empire of Alexander the Great fell to pieces after Alexander's death in 323 BC. The king's father, Mithridates V, had been poisoned by his wife as soon as she had a son who seemed likely to survive. She developed such a passion for poisoning that her son, alarmed by his mother's rapacity and cruelty, fled the kingdom.

Mithridates the Great, King of Pontus, seen on one of his coins. Obsessed with the idea of being invulnerable to poisoning, he experimented with both poisons and antidotes. He died by the sword in 63 BC. Well into modern times almost any supposed antidote to poisoning was known as a mithridatum.

He turned out to be quite as rapacious and cruel as his mother. He also had a justifiable fear of being poisoned himself. He eventually managed to kill his mother and all other relatives who might inherit the throne, married his sister Laodice, took over Pontus and at once began to expand its boundaries. He was very successful. Vastly energetic, he studied sciences and magic, collected art, spoke the dozens of languages current in Anatolia, was proficient in the use of all kinds of weapons, and gave prizes to the best writers of the day. He patronized mathematicians and philosophers. With his respected pharmacologist, Crataeus, he carried out his own researches into poisons and antidotes, using his many captives and prisoners as guinea pigs.

Once he controlled most of central Turkey, the Crimea, and large parts of Greece, Mithridates also had control over the developing Silk Road. The Romans became alarmed. Aware of their strength, he tried to reach accommodation with them. Politics in Rome were complex. Politics in Anatolia were equally difficult. Eventually Rome launched its armies against Mithridates. He won, slaughtering 80,000 of the enemy. One captured general, who had been especially greedy and devious, was killed by having molten gold poured in his mouth. A more modest one was set free, laden with gifts. Rome made a second attempt to subdue Mithridates, sending the general Lucullus against him. Again things went badly for the Romans, even though Lucullus returned to Rome in triumph, bearing the first cherry trees to reach Europe as part of his treasures. Mithridates, increasingly powerful and influential, sent embassies to both Rome and China. However, a third Roman force was sent and this saw his end. He was defeated and deserted by his allies, and his son. Realizing that there was no way out, he took poison.

He had had a theory that if he constantly took small doses of all the known poisons, together with a secret medicine, his body would learn to tolerate a mortal dose of any of them. Rats can indeed alter their physiologies in this way. Humans can't, but nevertheless the story runs that the strategem, or his regular doses of antidotes, worked: he was poisonproof and couldn't die. Aghast, he had to ask a nearby soldier to kill him with a sword. Pompey the Great, general in this finally successful war, found the Pontic king's formula. It was merely an unimpressive mix of rue, salt, nuts and figs. Perhaps this disappointing recipe was counter-propaganda. Even so, the idea of a substance to neutralize the effect of poisons took hold, and as there was increasing need for one in Rome, the formula was expanded. Galen's prescription for the emperor Marcus Aurelius eventually consisted of seventy-five substances, including viper's flesh, much cinnamon and a goodly amount of opium. He dissolved it all in the best wine that the imperial cellars had in order to make an alcohol extract that tasted good. In his *De Antidotis* (On Antidotes), he writes that Faustian Falernian wine had no equal. This he discovered by going through the Palatine cellars, beginning with wines at least twenty years old and tasting each vintage until he found the oldest one that still was sweetly drinkable. However his client was probably not much interested in the final taste: in his *Meditations*, Aurelius, a Stoic, reminds himself, 'Surely it is an excellent plan, when you are seated before delicacies and choice foods, to impress upon your imagination … that the Falernian wine is [merely] grape juice.' He took two doses of the mithridatum a day.

The viper's flesh was probably a theriac substance, in that Galen supposed that a regular intake of it might protect against a viper's bite. The same idea might have been behind Pliny's inclusion in the recipe of 'the blood of a duck found in a certain district of Pontus, which was supposed to live on poisonous food, and the blood of this duck was afterwards used in the preparation of the mithridatum, because it fed on poisonous plants and suffered no harm'. This is almost certainly a reference to the belief that hemlock is not poisonous to larks, quails or some sorts of duck. The birds which eat it do not excrete the poisonous alkaloids, and so become poisonous themselves.

In the ancient world, especially at the height of the Roman empire, almost anyone with any sort of power, or who existed anywhere near an imperial palace, took daily doses of mithridatum or theriac. Rivalries were intense, and poison an easy solution to conflict. Unexpected shifts of the imperial tiara were common. Nero, though he tried hard, could not poison his mother, Agrippina, supposedly because of the antidotes she took. Livia, terrifying wife of the Emperor Augustus, is supposed to have planned the route of the tiara after Augustus's death by poisoning off various side branches of the family. The ancient city contained for a while the notorious poisoners Apollodorus and Canidia, who, according to Horace, favoured hemlock in honey as the staple substance of their trade. Gaius (Caligula) collected poisons and killed gladiators, jockeys and horses in an attempt to manipulate the outcome of the contests on which he gambled.

Those powerful people who didn't trust the efficacy of mithridatums commonly employed official tasters or *praegustatores*, as most poisons were administered with food or drink. The people so put at risk were usually slaves or freedmen. They became so numerous that they formed a *collegium* with a *procurator praegustatorum* at their head. Tasters remained part of European court life for the next thousand years. They can't have been of much use, since most of the poisons employed against the powerful took an hour or two at least to take much effect. Some were cumulative. By the time the food tasted by the taster poisoned him, it would already have been consumed by the taster's employer. Perhaps tasters were status symbols.

If the great had great worries, amongst lesser folks power relationships could be equally bitter. Poison has long been the weapon of the greedy, the frustrated and the oppressed. At the end of the first century AD the satirist Juvenal claimed that poisoning had become an accepted way for wives to get rid of husbands and stepchildren, and for children to get rid of rich fathers or rival siblings. With little real medical knowledge, the general community sought scapegoats for unexplained illnesses or deaths, and often suspected poisoning when bacteria or viruses were the real culprits. Fear of mass poisoning was so strong that in 80 BC the dictator Sulla promulgated strict laws against it. Poisoning became, perhaps unjustly, associated with women. The first recorded instance of this was in 331 BC when an epidemic of some sort was blamed on a large number of women, who were accused of concerted mass poisoning. It is true, though, that poisoners took the advice of well-known elderly women, such as Canidia with her hemlock and honey. Tacitus wrote that one such notorious poisoner, Martina, was suspected when the emperor Tiberius's nephew Germanicus died under suspicious circumstances in AD 19. Germanicus had made an enemy of Plancina, wife of governor Piso, and Martina was her close friend. The poisoner was sent to Rome, where the Senate was investigating Germanicus's death. She died on the journey. Her body bore no signs of suicide, but a phial of poison was found hidden in a knot of her hair. Another female poisoner was the infamous Locusta. Convicted of many crimes during the emperor Claudius's reign, she was taken up by Agrippina, Claudius's second wife. Claudius died soon after. He was immediately succeeded by Nero, Agrippina's son. Nero supposedly asked Locusta to poison his younger half-brother, Britannicus. Once Britannicus lay dead, Nero suspended Locusta's death penalty and kept her as his adviser on poisons, organizing a school where she could train others in her art. She and her students were allowed to test her poisons on animals and convicted criminals.

The Greeks' and Romans' killing plants were not cultivated but brought to the cities from the wild. For the Greeks, the plants were collected by a group of people called *rhizotomoi*, which translates as 'rootcutters', or, more loosely, 'denizens of the hills and wild places'. These people took no part in the civilized urban culture of the great cities of Greece. Sometimes they belonged to separate ethnic groups such as the Archaeothebans, perhaps even more ancient inhabitants of the land. They were regarded with awe, fear and contempt. The Romans bought their herbs from a similar group, called the Marsi, or travelling people, another archaic race, this time from the hills of the Abruzzi. They, too, were regarded with both fear and contempt – although Galen admitted to consulting them about cures for snakebite – just as gypsy herbalists were well into modern times.

Like some gypsies, the *rhizotomoi* and Marsi used plants for their psychoactive properties. They were shamans and sorcerers as well as healers, Janus people as well as ones who sold Janus plants. When these alarming strangers came into the cities, selling their strange bundles of roots and leaves to apothecaries and doctors, any hawker among them was called either *agurtês*, 'the man who attracts a crowd', or *ochlagôgos*, meaning a 'seducer of the crowd'. The scenes that took place in the agoras of ancient Greece were directly comparable to the ones in more modern times, when hucksters and snake-oil men cried up their wares in the fairgrounds and village greens of Europe and America. Then, as now, there was widespread fraud and foolishness. One of Galen's important innovations was an insistence that doctors should know plants for themselves and not rely on these strangers selling strange things, or well-known things well adulterated.

With the collapse of the Roman empire, and the shift of medical and pharmacological research to Egypt and Arabia, and their renaissance under Islam, the darker side of these disciplines continued to develop too. There was an extraordinary burst of interest in alchemy and toxicology, the two becoming intimately entwined. Although the alchemists failed to achieve their goal of the transmutation of lesser metals into gold, they developed the chemistry of poisonous metals such as mercury, arsenic and phosphorus. One of the great manuals on toxicology is *Kitab ül-Sumum wa'l-Tiryaqat*, a work in five treatises, compiled by Ibn Wahashiya in the ninth century. It deals with plant, animal and mineral poisons. Written in Arabic, it is a compilation of Greek and Indian sources. Kings and princes hid it in their strongest and most secret treasuries, far away from warring wives and usurping children. They must have felt inviolable, for having read the work they knew how poisons could be detected by sight, touch, taste or the toxic symptoms which they cause. They knew to suspect poisoned drinks, toxic foods, dangerous clothes, lethal carpets, beds, skin lotions and eye salves. But they must too have become insane with suspicion. Another, equally important, book was *Poisons and their Antidotes* by the Arab alchemist Abu Musa Jabir ibn Hayyan. In its six chapters, mostly plant poisons are described, together with their modes of action, dosages, methods of administration, and the target organ which is attacked.

Rome, 1660. Night. A gilded sedan chair is being carried down a narrow medieval street. Ahead, two servants light the way with flaring *torcia*. When the little cavalcade reaches a small square on the slopes of the Velia hill, it stops and the *portantina* is put down. There is silence except for the splash from a tiny fountain. The *portantina*'s occupant is a woman. She peers from behind her curtains and sees crumbling façades, some with pieces of Roman marble half emerging from the plaster. At one of the houses, she watches another chair arrive. Its occupant knocks at the door of the house. A light appears as the door is opened, and a sumptuously dressed young woman slips in. A few moments later, the door opens slightly again, and a servant crosses the square to the sedan. Its occupant takes off a ring and hands it over. This is the third night that she has done this, and still she has not been admitted to the house. She has no idea what the ring is worth, but hopes at least that the stone is a good one. It is. She is being waved over to the doorway. Telling her servants to wait, her heart beating, she goes in.

The steps lead downwards, across a landing with fluted and carved walls, further down still, into a strange room with black and red walls. A brazier and candles glint on the silks and damasks as a group of women, some hiding their faces with masks, turn to look at her in the lamplight. From their midst comes an old but kindly-looking woman. She holds out her arms in welcome. The new arrival bursts into tears and throws herself, sobbing, into the arms of her prey. Soon, the stranger's collar and bodice have been loosened, revealing terrible wounds on her throat, shoulders and breasts. She is washed with the tender love of her sisters in trouble.

The old woman was Hieronyma Spara, later more simply nicknamed La Spara. She was by repute a witch and a fortune-teller. She also ran a school of poisoning in Rome.

In Europe, the poisoner's flora had been extremely stable for a thousand years. Then Arabic knowledge arrived, perhaps first in Venice, at some stage in the early sixteenth century, stimulating interest in ambitious Europeans. A new poison arrived too. In the East, it was already familiar in India and the Arab world. It was subtle and swift, and the death it produced horrible. The first European to describe it was the German Valerius Cordus, born in Hesse on 18 February 1515. His father and uncles were well-respected apothecaries, and he took a scientific interest in the substances that had made them wealthy. In his famous *Dispensatorium*, he mentions some curious foreign seeds that acted as an extraordinarily powerful emetic. He called these 'nux vomica'.

Like most exotic and dangerous drugs, nux vomica was first proposed as a cure for plague. Over the next four hundred years, as it expanded its range through Europe and North America, like so many other extremely poisonous plants it became used for a wider and wider range of ailments. In the late sixteenth century, it was still extremely rare. Gerard illustrates the seeds in his *Herball*, but he writes only that he knows not whether they come of herb or tree. He thinks, too, they may grow in the East Indies and was right. He quotes Avicenna, and writes: 'Of the physical vertues of the vomitting Nuts we think it not necessarie to write, because the danger is great, and not to be given inwardly, but mixed with other compositions, and that very curiously by the hands of a faithful Apothecarie.' Whatever British apothecaries thought of his warnings, the drug was taken up enthusiastically, first in Italy, then in France and Spain. It was soon found that the poison was cumulative, and large numbers of hardly noticeable doses eventually killed the taker. It was an ideal poison. Waves of poison hysteria took off throughout southern Europe in the late sixteenth century, lasting until the end of the following one. Like the contemporary witch hunts, the hysteria often centred on an elderly, gnarled woman, and commonly resulted in her death.

This appalling new 'medicinal' plant was the seed of *Strychnos nux-vomica*, a slight tree growing in India, especially along the Coromandel Coast, on the island of Ceylon, and in other parts of the East Indies. It is one of a number of circum-tropical species: some of the South American ones were used by inhabitants of the jungle to poison arrowheads. Here is a nineteenth-century account of nux vomica's operation:

In about an hour … the patient begins to feel uneasy from a sensation of impending suffocation. The tetanic convulsions then commence with great violence, nearly all the muscles of the body being affected at once. The limbs are thrown out, the hands are clenched, the head is jerked forwards and then bent backwards, and the whole body is perfectly stiff from the violence of the contractions. The pulse is very rapid; the temperature may rise. Hearing and sight are acute. The convulsion lasts a minute or two, then the muscles relax, and the patient feels exhausted and sweats all over. The intermission is short, convulsions soon come on again, and again there is a relapse to the state of muscular relaxation. The convulsions now rapidly increase in severity; and owing to the violent contractions of the muscles of the back, the patient is in the position of opisthotonos, resting on his head and his heels. The abdominal muscles are as hard as a board, the chest is fixed, the face becomes livid, the eyeballs are staring. The contraction of the muscles of the face causes a risus sardonicus but those of the jaw are not affected till quite the end. Consciousness is retained to the last. The slightest noise or even a bright light will reflexly bring on the convulsions, which may jerk the patient out of bed. Ultimately he dies from exhaustion and asphyxia …

By the mid-1600s, poisoning had become such an art that schools of alchemists, humanists and scientists occasionally taught how to poison clothing, gloves, jewels and letters, as well as the more traditional poisoning of food and drink. Overweening alchemists and scientists were sometimes killed by the Inquisition, but the schools were often tolerated. However, by 1659 it had come to the notice of Pope Alexander VII that great numbers of women, young and old, were confessing to his priests that they had poisoned their husbands with the new slow poisons. Even in the streets, it was popularly believed that young widows were unusually abundant, though no one seems to have noticed the yells of their poisoned spouses. The papal authorities had learned that one of the supposed poison schools met nightly at the ancient house of an old woman named Hieronyma Spara. Her pupils included women from amongst the greatest families of Rome and all learned about the virtues of nux vomica.

The new arrival at La Spara's 'coven' was a spy. She claimed to be locked in an extremely unpleasant marriage. The scars were her proof. Between her tears and the embraces of the others, she begged La Spara for a vial of her wonderful 'elixir'. La Spara detected no trap, but didn't yield the precious fluid for the next few meetings. Finally, she did, and the spy departed for ever. The contents of the vial were tested on animals, and found to contain poison. They died in a way that suggested the use of nux vomica. The old woman was arrested and tortured. She, an accomplice called La Gratiosa and three other women present on that fateful evening, and who were accused of poisoning their husbands, were hanged together at Rome. More of La Spara's pupils were later hanged or whipped through the streets.

A female poisoner who survived longer than La Spara was Madame Giulia Toffana (1653–1723), popularly called La Toffana. Based in Naples, she too had a dangerous recipe, for a concoction called 'aqua Toffana', which seems to have contained arsenic and belladonna. The phials that contained it bore a representation of St Nicholas of Bari, Bari being an ancient settlement of the Marsi people. The saint's tomb exuded a sacred liquid, which she claimed was what the phials contained. She sent her 'aqueta' (another name for aqua Toffana) all over Italy, and it is supposed to have killed about six hundred husbands.

La Toffana survived for so long partly because she pretended to be a person of great godliness. Whenever she was threatened, she retreated to the protection of a nunnery, and continued her trade from there. However, in 1723 she, too, was trapped at last. The

Strychnos nux-vomica, from *Phytographie Medicale* by Joseph Roques (1821). The large disc-shaped seeds of this tropical tree were a popular seventeenth-century poison, as they contain the alkaloid strychnine, which is still used against vermin. In minute doses, the seeds also found a wide range of medicinal uses, many now much frowned upon.

populace, who held her in high esteem, rioted to protect her. The authorities spread the idea that she had poisoned the city's wells. The populace turned against her. She was lost. Dragged from the convent where she had been hiding, she was tortured and strangled, and her corpse was thrown over the wall into the garden of the convent.

By then, the vogue for poisoning was solidly established elsewhere. In France, the most notorious poisoner was Catherine Deshayes Monvoisin, whose sinister nickname was La Voisin. Even her aristocratic clients could not stop her from being burned as a poisoner and a sorceress in 1680. The scandal caused the formation of a special court, the *chambre ardente* [burning court] to judge such cases. However, France's most sensational trial was centred not on the traditional witch figure of an old crone but on the young, beautiful Marie Madeleine d'Aubray, Marquise de Brinvilliers (*c.*1630–76). It renewed the focus on nux vomica. Marie's

The Marquise de Brinvilliers on the way to her execution, from a seventeenth-century painting by Charles le Brun. She and others inflicted terribly unpleasant deaths on their victims.

husband, Antoine Gobelin de Brinvilliers, older than she, must have decided early that he would be unable to keep her to himself. In the event, his complaisance saved his life. His wife came under the influence of a treacherous army captain, Godin de Sainte-Croix, who became her lover. It was Marie's father, not her husband, who was incensed at such behaviour and tried to terminate the scandalous relationship. That was a mistake. He had Sainte-Croix thrown into jail, but there the prisoner met a professional poisoner supposedly once in the employment of the ex-Queen Christina of Sweden. Mysteriously called only Exili, he knew all about nux vomica.

Once out of prison, Sainte-Croix, with his insatiable appetite for both Marie and money, persuaded Marie to poison her father, two brothers and a sister in order to secure for them both the family fortune and to end interference in their relationship. First they needed to do some experiments. Marie began to make charitable rounds of the city's hospitals. Once she had finished testing dosages of nux vomica on the patients, she poisoned her father. Her brothers inherited. They, too, soon died in torment. Embarrassed by a previous liaison, she attempted to poison her children's tutor, a M. Briancourt. He fled, and also ensured that the Marquise's sister-in-law and sister vanished into the safety of a convent. Marie was becoming exceedingly dangerous. Bored by her daughter's stupidity, she poisoned her too, though she apparently regretted it immediately afterwards and made her drink a great quantity of milk. But her father and brothers had died in such agony that suspicions had been aroused. An investigation was made, and the Marquise fled abroad without Sainte-Croix. He then died an untimely (but natural) death in Paris. Amongst his papers were found documents that made the Marquise's guilt plain. After several years on the run in England and the Netherlands, Marie was arrested at Liège in 1676. She was tried and convicted on all charges. Terribly tortured, she was beheaded on the scaffold in the Place de Grève, and then burned. Thousands watched.

During the eighteenth century, nux vomica, now called an 'inheritance powder', moved to Britain. There it was at the centre of one of the early nineteenth century's most sensational poisoning cases. Thomas Griffiths Wainewright (1794–1847) was orphaned early, and brought up by his grandfather, editor of *The Monthly Review*, London's first literary magazine. His grandfather's lifestyle – he moved in a world of writers, dandies, dilettantes, painters and poets – determined various aspects of Wainewright's future. He became a fairly successful painter, but he had a far more dangerous passion: to be a gentleman. At thirty, he was at the centre of artistic circles in London, was a friend of William Blake and Charles Dickens, and was exhibiting paintings at the Royal Academy in Piccadilly. At forty he was on a chain gang in Tasmania, greed and nux vomica having been his undoing.

By 1822, with a young wife, Eliza, and an extravagant lifestyle, Wainewright was desperate for money. His grandfather was by now dead, his money in a trust fund. Wainewright turned to forgery. Eventually, he managed to get hold of his inheritance. It didn't last long, and he soon had huge debts. An uncle decided to help, inviting Wainewright, his wife and their son to move into his late grandfather's Linden House. The uncle soon died in mysterious circumstances, leaving his house and estate to Wainewright. The

Head of a Convict, very
characteristic of low cunning &
& revenge!

Thomas Griffiths Wainewright's obsessive need to be seen as a
gentleman led to his extensive use of 'inheritance powder' upon members
of his family. In this self-portrait of the 1840s his morbid stare was at the
landscapes of Hobart, Tasmania, after his deportation.

Wainewrights then invited Eliza's mother and two sisters to come and stay with them. Money was still a problem. Wainewright, Eliza and Helen Abercrombie, the youngest sister, set up an elaborate insurance fraud in 1829, insuring Helen's life with five different companies on false pretences, lying about Helen's age and the Wainewrights' financial situation. The girls' mother was desperately alarmed, but soon died in mysterious circumstances. Helen, a healthy twenty-one-year-old, was too trusting. As soon as the documents were signed, she became ill. A few days later, in the grip of awful convulsions, she died. A servant reported that her death was identical to those of Wainewright's uncle and Mrs Abercrombie.

The insurance companies refused to pay out on the basis of 'misrepresentation'. Wainewright fled to France. No charge of poisoning was ever brought against him, though several books on poisons were found in his library. Life in France was hard, and Wainewright returned to London. He was arrested almost immediately, held in Newgate prison, and then shipped off to Hobart. In Newgate, he was reported by a fellow inmate to have said, as if a character in a Dickens novel: 'I will tell you one thing in which I have succeeded to the last. I have been determined through my life to hold the position of a gentleman. I have always done so. I do so still. It is the custom of this place that each of the inmates of a cell shall take his morning's turn of sweeping it out. I occupy a cell with a bricklayer and a sweep, but they never offer me the broom!' The paintings he did when at Hobart are generally thought to be his best work as an artist.

Although it was used in Europe as a poison, nux vomica's future lay in nineteenth-century America. There, doctors had become more and more obsessed with the use of small doses of extreme poisons as cures for practically everything. No doubt the slight effect of a small dose of bitter nux stimulated the placebo effect. One doctor, Dr Felter, wrote: 'Were I restricted to five remedies for the alleviation of human ills, Specific Medicine Nux Vomica would hold a prominent place among the five.' He, and many like him, used it for children's colic, and for breathing difficulties of all sorts from apnoea to the panting breath of diphtheria. It was used as a laxative. It was used to ease labour pains, to rouse frigid women to passion and for assorted other 'women's difficulties'. It appeared in numerous quack medicines, and was easily bought over the counter for self- or spouse-dosing. There were intense debates about whether it was better to use the new extract from the nuts, called strychnine, than the whole seed powder, called nux vomica. Some doctors, sensibly, broke away from the usage of either, noticing, as had La Spara, that the body only slowly excreted the alkaloid. Dr Felter later goes on:

... the actual poisonous effects of Strychnine were to be seen in many directions where physicians thoughtlessly gave small doses of Strychnine until its cumulative lethal effect followed ... Especially is it to be deprecated that the laity have learned to use these ready-made tablets and compounded pills of Strychnine as a constant self-treatment as a laxative. When persons without discriminative care or professional knowledge turn to the bottle of 'Strychnine, Belladonna and Aloin,' or similar pellets, as if they were innocuous substances ... it is as a child playing with fire over a barrel of gunpowder.

Nux vomica is still available today, and recently reappeared in America, mixed with yohimbine and testosterone, as a 'natural' way of giving a reluctant penis an erection. Terrifying.

The Dr Jekyll-ish debate among late-nineteenth-century American doctors as to whether extreme poisons should be used in medicine had major repercussions in the subsequent history of the search for health. On the one hand, dilutions of poisons became more and more extreme, until almost nothing, or truly nothing, was left. Practitioners still got results, and homeopathy was born. Debate still rages over whether homeopathy invokes anything more than the placebo effect. On the other hand, the discussion of whether to use extracted alkaloid or a preparation based on the whole plant gave rise to various stoutly defended schools of thought. Some, such as Samuel Thomson, rejected the use of conventional plant poisons altogether. It later turned out that the plant whose use he promoted was almost as poisonous as the ones for which he intended it as a substitute.

Thomson was a man of obsession, and inhabited an ambiguous mental region, not uncommon in 'healing' circles, where quackish ideas are promoted with profound conviction. He was born on 9 February 1769, in the town of Alstead, Cheshire, in New Hampshire. He wrote:

> There was an old lady by the name of Benton lived near us, who used to attend our family when there was any sickness. At that time there was no such thing as a Doctor known among us, there not being any within ten miles. The whole of her practice was with roots and herbs, applied to the patient, or given in hot drinks, to produce sweating; which always answered the purpose. When one thing did not produce the desired effect, she would try something else, till they were relieved.

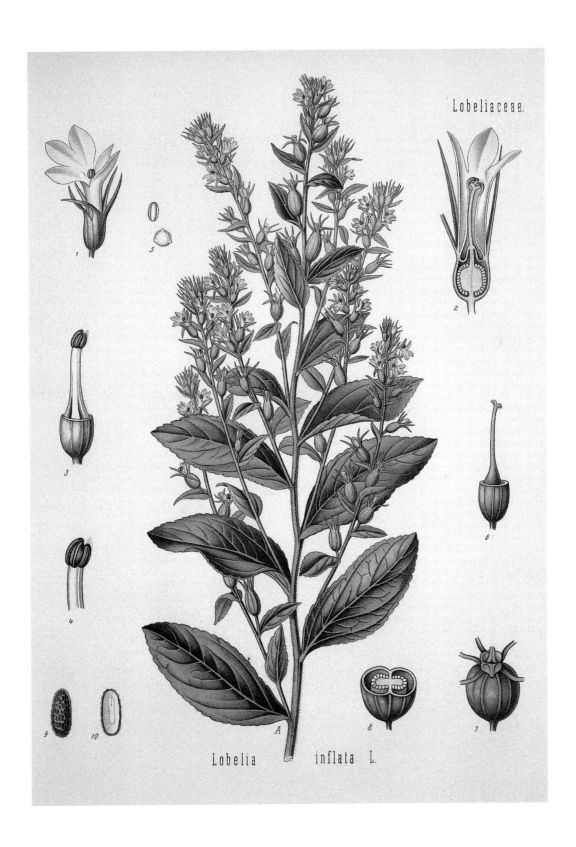

Lobeliaceae.

Lobelia inflata L.

He implies that she chose her plants out of the local flora almost at random and, following her lead, he also soon discovered the plant on which he eventually took out a patent. It had 'a singular branch and pods, that I had never before seen,' he wrote, 'and I had the curiosity to pick some of the pods and chew them; the taste and operation produced was so remarkable, that I never forgot it. I afterwards used to induce other boys to chew it, merely by way of sport, to see them vomit.' He was to play with it in this way for twenty years, gradually building a belief system around it: 'It is a certain counter poison, having never been known to fail to counteract the effects of the most deadly poison, even when taken in large quantities for self-destruction.'

He had become interested in medicine after a horrible incident in his early life – which also shows how tough life was for settlers with no access to doctors and no knowledge of how to treat medical crises. 'In the year 1788, when I was in my nineteenth year, my father purchased a piece of land on Onion river, in the state of Vermont, and on the twelfth day of October, he started from Alstead, and took me with him, to go to work on the land and clear up some of it to build a house on, as it was all covered with wood.' There was an accident. Samuel's foot was wounded, and the wound festered. 'My father in dressing my wound had drawn a string through between the heel-cord and bone, and another between that and the skin; so that two-thirds of the way round my ancle was hollow.' The wound blackened, and they had to set off through the virgin forest with Samuel tied to a plank dragged by the only horse. After a hair-raising journey, he found succour at the hands of the gruff Dr Kitteridge. Samuel's leg was saved, and he determined to doctor. Thomson remembered:

The herb which I had discovered when four years old, I had often met with; but it had never occurred to me that it was of any value as medicine, until about this time, when mowing in the field with a number of men, one day, I cut a sprig of it, and gave it to the man next to me, who ate it; when he had got to the end of the piece, which was about six rods, he said that he believed what I had given him would kill him, for he never felt so in his life. I looked at him and saw that he was in a most profuse perspiration, being as wet all over as he could be; he trembled very much, and there was no more color in him than a corpse. I told him to go to the spring and drink some water; he attempted to go, and got as far as the wall, but was unable to get over it, and laid down on the ground and vomited several times. He said he thought he threw off his stomach two quarts. I then helped him into the house, and in about two hours he ate a very hearty dinner, and in the afternoon was able to do a good half day's labor. He afterwards told me that … he felt better than he had for a long time. This circumstance gave me the first idea of the medical virtues of this valuable plant, which I have since found by forty years' experience, in which time I have made use of it in every disease I have met with, to great advantage, that it is a discovery of the greatest importance.

The plant seems to have been *Lobelia inflata*. Other doctors pointed out that it was a poison almost as strong as nux vomica. Endless litigation ensued. The publicity and ridicule did not stop him from recruiting adherents, but fortunately the plant did not develop widespread use.

While we humans have used plants to poison members of our own species, we have also explored the world's flora in an effort to find plants that will poison other animal species in the world around us. We have found plants that can poison the fish in rivers, kill the birds that eat our grain, and plants that rid our crops of the insects that gnaw or suck the life from them. We have found plants to poison the carnivores that hunt our livestock or compete with us in the hunt – we have the 'bane' plants, such as dog's bane, wolf's bane, leopard's bane and so on, once used to smear on bait. We have poisons to tip spears and arrows, sometimes with such potency that the blade needs only nick the skin of the attacker. Every culture has found such plants.

In Europe, one of the oldest and also most familiar of these is the common yew (*Taxus baccata*). The sticky juice from beneath the bark was once a formidable arrow poison. Pliny mentions that arrows of Scythian tribes were smeared in yew sap to make them poisonous. The only problem with yew was that a yew-juice-filled wound took two hours to affect the prey and hunters needed to be fit to cope with an animal such as a deer while it lost its footing, started breathing shallowly, vomited, before eventually collapsing with respiratory paralysis.

The Scythians, a nomadic and warlike tribe of expert horsemen, tipped their arrows and spears with the sticky poisonous sap of the yew tree. They were also ardent users of *Cannabis sativa*. The archer here is seen on a Greek vase of 570 BC.

In the tropics there were plants with faster-acting juices. Other members of nux vomica's genus, *Strychnos*, were and are particularly rich in poison. Arrows poisoned with the Javanese *Strychnos tieute*, a climbing shrub, caused death by violent convulsions, the animal's heart stopping before its lungs collapsed. The South American *S. toxifera* gives the deadly curare (the word is a corruption of *woorari* or *urari*). When the natives of British Guiana use this poison, a wounded animal needs tracking for often much less than twenty minutes.

Today, the ancient search for arrow poisons has also given a few clues as to what might work for the pressing needs of farmers and gardeners. Since many of the modern insecticides produced by the chemical industry have proved to be very damaging, being slow to decay and dangerous to wide ranges of animal types, 'botanical' insecticides, which decay rapidly and sometimes poison quite specific ranges of insects, have become increasingly important. One that is particularly important at the moment is derris, the powdered bark of a number of species of *Derris*, twining members of the *Leguminosae* or pea family that grow along wooded river banks in tropical Asia and South America. The most common species, *Derris elliptica*, is an ancient arrow and fish poison, which allows hunters to harvest all the fish from large stretches of water. The plant's main component, rotenone, is rapidly detoxified in the human gut, so the fish are entirely edible. Commercial growers like derris because it kills insects and a crop can be marketed soon after spraying. However, rotenone kills all insects, beneficial predators and parasites as well as the pest species.

Various tanacetums, including the herbaceous red or pink *Tanacetum coccineum* familiar in our gardens, yield another popular insecticide. *T. cinerariafolium*, in particular, is widely farmed for its pyrethrum. This substance rapidly kills aphids and caterpillars. It also kills beneficial arthropod predators such as lacewings, hoverflies and ladybird larvae. However, as it decays rapidly in air, vanishing within twelve hours, plants sprayed in the evening will not poison bees alighting on them the following morning. It is one of the oldest and safest insecticides available. The pyrethrum paralyses insects almost immediately, to spectacular effect. Many of the immobilized insects later recover. Pyrethrum is scarcely toxic to mammals.

Some plant insecticides are more narrowly toxic. One such is quassia, made from the bark and root of a South American tree called *Picrasma excelsa*. A spray is simply made by boiling wood chips in water, straining the liquid through a filter and adding a wetting agent. It is an effective killer of aphids and small caterpillars, but doesn't poison ladybirds and ladybird larvae, though it does kill hoverfly larvae. It is widely used. An insecticide can be extracted too from the seeds and leaves of the neem tree (*Azadirachta indica*), and its effects have been known in India for thousands of years. The tree's juices disrupt insect growth and the ability of the females to lay eggs. Neem insecticides are most effective on insect nymphs or larvae, especially those of moths.

The tobacco, here in a drawing from *Hortus Eystettensis*. One of many mind- and mood-altering plants exploited by native Americans, tobacco has powerful insecticidal properties too.

The insecticide in longest use by Western gardeners is nicotine. Leaves of the tobacco plant, *Nicotiana tabacum*, were smoked by the pre-Columbian natives of North America, and the addiction to it spread to Europe almost as soon as America was discovered. Native Americans also sometimes used a maceration of the leaves as an insecticide for their potatoes, maize, tomatoes and other staples. In Europe, it was 1690 before gardeners latched on to the idea. Previously, plants under attack by insect were sprayed with hot water, or the insects were

killed off by hand. The new substance caught on, though raw tobacco juice was soon discovered to be extremely toxic to mammals as well as to caterpillars, aphids and earthworms. In mammals, nicotine, the principal alkaloid in tobacco, mimics a substance that occurs at the junction of muscle and nerve. A poisoned creature starts to twitch, starts having convulsions and dies soon after. In insects, nicotine attacks the junctions between nerve endings in the central nervous system. The alkaloid persists in the environment for up to a week, so crop plants that have been dusted with a nicotine powder, or sprayed with a nicotine liquid, cannot be eaten for several days. It is by far the most hazardous of 'botanical' insecticides.

Over the last half century, tobacco, when smoked, has become associated with numerous human health problems, especially lung cancer. Other plants, even ones that have long histories of medicinal use, can also be powerful carcinogens, either when ingested or by skin contact. The list of plants that can kill us slowly and secretly is growing by the year and includes plants like comfrey (*Symphytum* spp.), ragwort (*Senecio jacobaea*) and tansy (*Tanacetum vulgare*), pokeweed (*Phytolacca americana*), hemlock (*Conium maculatum*), *Lupinus* spp., eupatorium (*E. purpureum*) and senna species; St John's wort (*Hypericum perforatum*) has just joined the list.

There is, though, a rather deeply tarnished silver lining to this story. Other, intensely poisonous, plants are turning out to be surprising healers of cancer. Some plants can, when used with great care, poison cancer cells before they damage the whole human. An early discovery was one of the many exotic substances contained in the pretty Madagascar periwinkle (*Catharanthus roseus*). The plant has often been used to induce euphoria, and as an aphrodisiac. An excess of the plant can easily cause major damage to the brain, but one of the substances contained in its glossy leaves, called vincristine because most periwinkles belong to the genus *Vinca*, was found to kill the cells involved in some sorts of brain cancer before it killed normal brain tissue. Vincristine is still in use, though it has so many appalling side effects that it is now used in short bursts, alternated with other drugs. Ever ambiguous, in combination with other drugs, it can also act as a powerful carcinogen.

As well as an arrow poison, yew, that tree of churchyards and silent forests, also yields an increasingly important anti-cancer drug. When, in the early 1960s, American cancer specialists began looking for anti-cancer drugs in native plants, they found in the bark of the Pacific yew tree (*Taxus brevifolia*) something that stopped tumour growth in a wide variety of rat tissues. The substance was named taxol. Gradually its unique mode of operation, by which it stops cancer cells dividing, became clear. However, the Pacific yew is rare and very slow-growing. Six 100-year-old trees could just about provide enough taxol to treat one patient. Other species of yew were examined. The European and Asian yew, *Taxus baccata*, turned out to contain a closely related analogue of taxol called baccatin III. Turning this into taxol was a slow and difficult process, but at least there were plenty of *T. baccata* available. Or so it seemed. By 1989, in spite of its wide range of side effects, taxol was being used against advanced ovarian cancer and in 1994 its use was approved for breast cancer. Pharmaceutical companies soon stripped the Indian forests bare. Gardeners came to the rescue. In many gardens, lawns and flower beds are flanked by high dark yew hedges that need clipping several times a season. The clippings now have a good market value. Many owners of yew hedges bag them and send them off to companies who can extract their taxol for the benefit of humankind.

Catharanthus roseus, shown in an engraving by Sydenham Teast Edwards from Curtis's *Botanical Magazine*. This subtropical periwinkle is the source of vincristine, a substance used in the treatment of some sorts of brain cancer, Kaposi's sarcoma and Hodgkin's lymphoma. Though also used in the tropics as an aphrodisiac, it is extremely toxic.

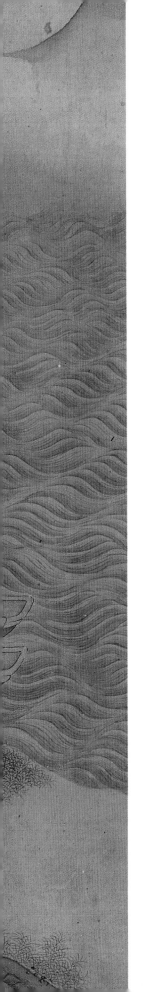

THE SEVEN AGES OF MAN

THE BAND appears. Ooom pah! Oooom pah! Oooom pah! Flocks of startled pigeons burst up into the sunshine. Dogs bark. Cats slink away. The bandsmen march swiftly up the street, uncomfortable in their ill-fitting tattered pink-braided jackets. Children shriek. Parents stare. Through the town square, past saloons and drugstores, the gaol, the courthouse, churches, and shuttered and balconied houses, music echoes. Following the band are seven hastily commandeered carts, decorated in tinsel and ribbons. In each cart sits a woman. None is young. None is handsome. All are well dressed. All have long dark glistening hair which, with long-practised coquetry, they display to the crowd. The carts are decorated with boards, proclaiming slogans such as 'A woman's hair is her crowning glory' and 'Remember, ladies, it is the hair not the hat that makes you beautiful.'

The year is 1893. The women are, or rather were, the Seven Sutherland Sisters on a promotional trip out of town. One of the sisters, Naomi, and her bass voice, has just died. This has been a great difficulty. Their father was a rogue. Once a Methodist minister, he became a politician, but that was a disaster. Casting round for ideas, he realized that his daughters, all musical, were an asset. He put them on the stage as a vaudeville troupe and set himself up as their promoter. This did better. They were taken up by the Barnum and Bailey travelling circus – the 'Greatest Show on Earth'. The girls enjoyed themselves. Mr Sutherland soon noticed that their long tresses caused rather more of a stir than their songs. Their long-deceased mother, though parentage was

The first Chinese emperor, Ch'in Shi Huang Ti, having united warring states and initiated the building of the Great Wall, became obsessed with the need to live for ever. He travelled his immense territories hunting for plants that would prove to be his elixir, and thereby met his end.

All seven Sutherland sisters, at the height of their commercial careers, showing off their abundant torrents of hair. Their roguish father reclines in the foreground. The Sutherland 'Hair Grower', later analysed, contained no exotic substances whatever.

suspect for one daughter and one son, had anointed her children's hair with a strong-smelling concoction of her own devising. It may have been something to deter head lice, but her daughters maintained that it gave them remarkable heads of hair. They said that the mixture was an old family secret. Their father had an idea. In 1884, he concocted the 'Seven Sutherland Sisters Hair Grower', and marketed it after each performance.

More of a commercial hit than his daughters, the 'Hair Grower' was soon making a substantial income. All over America, hopeful women began anointing themselves with the dark liquid in the hope of having immense and alluring tresses. They ended up paying the Sutherlands $90,000 in the product's first year of sale. As the sisters were its main promotional gambit, when one of the sisters, Naomi, died, the family panicked. They wondered whether to turn the dead Naomi's hair into a wig. However, they were saved from anything so bizarre by the discovery of Anne Louise Roberts, who had her own nine-foot-long tresses. She passed herself off as Naomi for a number of years, though presumably she didn't have Naomi's bass voice as well. Naomi was expensively buried with her own hair.

The 'Hair Grower' and its related products, for there were soon hair dyes too, sold well into the twentieth century. Yet the large fortune their father's opportune ideas produced gave most of the sisters no lasting wealth. Almost all became both obsessed by their own hair, and exceedingly extravagant. One was locked away in an asylum. Another died in extreme poverty. Only one married (she became Mrs Bailey). When the secret formula that sealed their fates was analysed, it was discovered that the bottles contained little more than alcohol, quinine and borax. At least it couldn't have done any great harm. Its great rival, the confusingly named 'Princess brand Hair Restorer and Bust Developer' went one better and offered a money-back guarantee if any harm were done.

Effective or not, the Sutherland products appealed to women because of the fundamental human desire to be attractive to a mate and to reproduce. In all the aspects of women's lives relating to procreation – menstruation, conception, growth or abortion of the foetus, giving birth, lactation, the cessation of fertility, all involving remarkable waves of hormonal change – an immense number of plant species have been or still are used, in the belief that they affect the processes involved. Some plant uses have hardly appeared in herbals, existing for centuries as the half-secret knowledge of midwives.

Plants were used to increase the menstrual flow. Plants were used to decrease it. Even now, plants are being exploited as a means of regulating the emotional impact of the period. The ubiquitous feverfew (*Tanacetum parthenium*) is one of the oldest plants associated with the period, whether its pains and cramps, or what was called 'sluggish flow'. Ancient Greek women took it to relieve menstrual discomfort, and the use persisted into sixteenth-century Europe. No doubt that was one of the reasons why European settlers took the plant to the New World. It is now found along roadsides and wood borders from Quebec to Ohio and south to Maryland and Missouri, and westwards to California. The hop plant (*Humulus lupulus*), a rampant climber native to central and southern Europe and western Asia, which provides a spring vegetable and a flavouring for beer, was used because it 'bringeth down women's courses'. Modern research has found some as yet unidentified oestrogenic substances in the plant. These seem to have no effect on the menstrual cycle (except that the mild sedatives contained in the plant's leafy and resinous seed heads may affect the emotional aspects of the menses), though they can produce mammary development in men. Hop is related to cannabis, which, like the poppy, was also called into action, at least from the seventeenth century. As late as 1890, Queen Victoria's personal physician, Sir Russell Reynolds, prescribed cannabis for her menstrual cramps. He wrote in the first issue of the medical journal *The Lancet*: 'When pure and administered carefully, [cannabis] is one of the most valuable medicines we possess.'

Dioscorides suggested the milky juice of 'Panax Heraklios' (probably *Ferula gummosa*), for menstrual cramps and labour pains. Others suggested willow bark, whose salicylic acid would have alleviated period pain. Very much more dangerous was the use of a suppository of belladonna (*Atropa belladonna*) for the same difficulty. For seventeenth-century women who had painful periods Gerard suggested rue, rosemary, an umbellifer which was most probably *Anthriscus sylvestris*, and the dangerous dittany (*Dictamnus*). Recent work on *Anthriscus sylvestris* has shown that, at least in laboratory animals, its terpenoids can indeed profoundly affect the hormonal system, preventing ovum implantation, inhibiting ovarian growth and disrupting the oestrus cycle. Its effects on human females have not been studied. It is mentioned in an ancient children's skipping rhyme, though the obscure wording may mean that it was used as an abortifacient rather than to reduce menstrual pain:

Queen Anne's lace
Saves women's place.
I love that flower
To keep my power.

In China, where menstrual difficulties were attributed to imbalances of various forces, doctors prescribed the usual wide-ranging mix of plant substances. One plant used on its own was the Chinese foxglove (*Rehmannia glutinosa*). Used for irregular menstruation, uterine and postpartum bleeding, it was also supposed to cure men's nocturnal emissions. Chinese angelica (*Angelica sinensis*) was believed to have similar properties.

A plant recently much touted for these purposes was the evening primrose (*Oenothera biennis*). As well as being edible, used in pickles, soups and sautés, and looking good in the garden, it produces seeds that are rich in oils containing a fatty acid called gamma linolenic acid. It also contains unknown anti-coagulant substances. The oil squeezed from the seeds is recommended for a wide variety of ailments, including difficulties associated with periods and menopause. Some enthusiasts also promote it as an easy slimming aid, one that happily enables those who take it to achieve weight loss without dieting. It is also used for hypertension, rheumatoid arthritis, multiple sclerosis and even hangovers. Gamma linolenic acid can indeed be converted to a hormone-like compound called prostaglandin E1, a substance which may increase blood flow, decrease blood clotting and reduce inflammation. These effects have not yet been confirmed. There are now some doubts about the safety of long-term use. It certainly increases the risk of temporal lobe epilepsy in people with schizophrenia.

The American flora also has something else really much more important – for female and other problems. Some native American tribes of the north-east, the Delaware and the Cherokee, used a plant called cohosh to promote menstrual flow. Other tribes used it in more general ways as a purgative, a stimulant, to calm fevers and to drive worms from the gut. Yet others used it for diarrhoea, snakebites, sore throats and rheumatism. With such a huge cure list, it is not surprising that it was one of the first native herbs taken up by European settlers, such as North America's first apothecary, Louis Herbert (1575–1627).

It took a while for black cohosh (*Actaea racemosa*, syn. *Cimicifuga racemosa*) to catch on in Europe. It was first described by French botanists in 1705. By 1732, it had been introduced into English gardens as a hardy ornamental perennial, where it has remained modestly popular ever since. No one in Europe explored it as a medicinal plant, though it continued to be used in the developing America of the eighteenth century. It appears in the first work on American medicinal plants, published in 1801, and became increasingly important during the rest of the century. It was widely prescribed by physicians for normalizing difficult menses and to relieve pain after childbirth, and also as an anti-inflammatory for arthritis and rheumatism. It was listed as an official drug of the United States Pharmacopeia from 1820 to 1926. The reason it was dropped was probably that it became part of the recipe of one of the most famous 'quack' medicines of all time: Lydia Pinkham's Vegetable Compound.

Interest in native American medicine started as soon as European settlers moved away from early colonies and from European doctors and European cures. French apothecary Louis Herbert began collecting Canadian tribes' medicinal plants in the early seventeenth century. Seqoyah (*c.* 1760/70–1843) belonged to the Cherokee, a tribe who used black cohosh to treat their womenfolk.

Lydia Estes was born in 1819, the tenth child of Rebecca and William Estes of Salem, Massachusetts. She grew up into an energetic young woman who, even early in her life, managed an interesting mix of high-mindedness and quackery. Though she eventually became a schoolteacher in her home town, she often claimed to be a trained midwife and nurse. She joined the Female Anti-Slavery Society. She promoted temperance. She studied phrenology, spiritualism, and various forms of evangelism. In 1843, she made a good marriage to the wealthy property owner Isaac Pinkham. She soon started a family. Life went well. She became famous, and eventually enormously rich, only when her husband's real estate investment fell to pieces after the upheavals of 1873. Faced with severe economic hardship, but not easily defeated, Lydia turned to medicine, and began brewing potions on the kitchen stove at her small house, 271 Western Avenue, Lynn, Massachusetts (which still exists). No slouch at self-promotion, she called her brew 'the greatest medicine since the dawn of mankind'. Selling for a dollar a bottle, her 'Vegetable Compound' was touted as a mighty elixir for all nineteenth-century 'woman problems'. Advertisements, which suggested that 'Some of the conditions which disappoint the hope of children are displacement of the womb, constriction of the ovaries, local catarrhal conditions, obstructed menstruation, and abnormal growth of tumors', proposed a daily dose from the bottle. Much of her pitch was based on a distrust of the medical profession: she advised her customers to 'let the doctors alone'; they would no longer be needed in the face of her 'positive cure for all these painful Complaints and Weaknesses so common to our best female population' that would 'cure entirely all Ovarian troubles, Inflammation and Ulceration, Falling and Displacements, and any consequent Spinal Weakness, and is particularly adapted to the Change of Life'. Lydia's portrait gazed upon womanhood from every bottle. She published articles and answered letters in ladies' magazines and free pamphlets, giving out advice that always suggested a dose of 'Lydia E. Pinkham's Vegetable Compound'.

Women began buying this new product cooked up by a woman for women's problems. Soon the demand was so high that Lydia set up 'the Factory' near her home in Lynn, at 305 Western Avenue. At first she employed only women workers. She also employed other women to write as agony aunts for every newspaper and magazine she could persuade to take their copy with its aggressive promotion of her 'compound'. These advertorials were often called 'Facts and Fancies' or 'Lydia E. Pinkham's Private Textbook Upon Ailments Peculiar to Women'. All included lavish testimonials. One read thus:

> *Eight years ago I got into an awful condition with what the doctor said was falling of the womb. I would have spells of bearing-down pains until he would have to give me morphine, and when I could not stand that they would put hot cloths to me. The doctor said I would never have any children without an operation. A neighbor, who knew what your medicine would do, allowed me to give Lydia E. Pinkham's Vegetable Compound a trial. I did so and I have never had a return of my old trouble. The next September I gave birth to as healthy a boy as you can find, and now I have two more children.*

Such encomiums gave rise to the slightly sinister slogan 'A Baby in Every Bottle'.

She expanded the range, and soon the fabled Vegetable Compound was accompanied by a blood purifier, liver pills and a cleansing wash. According to one version of 'Lydia E. Pinkham's Private Textbook Upon Ailments Peculiar to Women', the maladies that could be treated effectively by Pinkham products ranged from anaemia to rheumatism and included kidney diseases, colds, impure blood, mental derangement, hysteria and nervous disease, sterility, tumours and dyspepsia.

The ingredients in Lydia's compound varied over the decades, but were mostly assembled from liquorice, chamomile, pleurisy root (*Asclepias tuberosa*), Jamaica dogwood (*Piscidia piscipula*), black cohosh, life plant (possibly a species of *Kalanchoe*), fenugreek seed and dandelion root. All were extracted in alcohol, which made up about 18 per cent of the 14½oz bottles. She explained to medical authorities and the Women's Christian Temperance League that the alcohol was only there as a solvent and preservative. Nevertheless, a tablespoonful or more would have been pleasantly relaxing, whatever the effect of the other ingredients.

Soon, her medicine was being sold worldwide and her name and picture were famous. At its peak in 1925 her business grossed $3.8 million. A few years later, the powerful American Medical Association was growing restive. An editor wrote: 'Another day, another ingredient, but still essentially the same old female weakness nostrum. Grandma used it, her daughter tolerated it, but her granddaughter should know better.' He clearly believed that the malleable recipe of the compound made it a 'quack' medicine. Indeed, most of its contents were useless. But black cohosh is different.

The Pinkham papers, a fascinating repository of purchasers' queries, testimonials and so on, are now ensconced in the Schlesinger Library in Cambridge, Massachusetts. The name is owned by an American herbal company. Whether Mrs Pinkham was crooked or straight, it is worth noting that in 1876, the year that she patented her Vegetable Compound, a prominent American physician was urging the removal of healthy ovaries as a treatment for menstrual cramps. The mortality for that operation was 40 per cent.

However, she wasn't the only person to advocate the virtues of black cohosh. Among them was Dr John King (1813–93). As lobelia was to Thomson, and nux vomica was to

LYDIA E. PINKHAM'S
VEGETABLE COMPOUND
IS A POSITIVE CURE
For all those painful Complaints and Weaknesses so common to our best female population.

It will cure entirely the worst form of Female Complaints, all Ovarian troubles, Inflammation and Ulceration, Falling and Displacements, and the consequent Spinal Weakness, and is particularly adapted to the Change of Life.

It will dissolve and expel tumors from the uterus in an early stage of development. The tendency to cancerous humors there is checked very speedily by its use.

It removes faintness, flatulency, destroys all craving for stimulants, and relieves weakness of the stomach. It cures Bloating, Headaches, Nervous Prostration, General Debility, Sleeplessness, Depression, and Indigestion.

That feeling of bearing down, causing pain, weight, and backache, is always permanently cured by its use.

It will at all times and under all circumstances act in harmony with the laws that govern the female system.

For the cure of Kidney Complaints of either sex, this Compound is unsurpassed.

LYDIA E. PINKHAM'S VEGETABLE COMPOUND is prepared at 233 and 235 Western Avenue, Lynn, Mass. Price, $1.00. Six bottles for $5.00. Sent by mail in the form of pills, also in the form of lozenges, on receipt of price, $1.00 per box, for either. Send for pamphlet. Address as above.

No family should be without *LYDIA E. PINKHAM'S LIVER PILLS.* They cure constipation, biliousness, and torpidity of the liver. 25c. per box.

Sold by all Druggists.

COMPLIMENTS OF

TRYON & CO..

DRUGGISTS,

Cor. Porter Av. and 13th St.   BUFFALO, N. Y.

TOP LEFT: Lydia Pinkham, whose Vegetable Compound contained plant constituents in large amounts of alcohol. TOP RIGHT: An early and encouraging advertisement for some of her products. BELOW: This tasteful, if not botanically accurate, packaging anticipated modern marketing of medicinal plants. The text weaves in a suitably alcoholic manner.

various prominent doctors, cohosh was for King, Professor of Obstetrics at a medical college in Cincinnati, his 'favorite remedy' (though he was also an enthusiast for echinacea). Following native American usage, he used it for acute and chronic cases of rheumatism and other inflammatory conditions, various lung and nervous afflictions, and 'in abnormal conditions of the principal organs of reproduction in the female'. King's later publications had enormous influence in Germany, and, like echinacea, black cohosh was taken up by German doctors. It is still widely used in that country. It is in Germany, too, that most modern research into its effects is taking place. At present, it seems as if the American Indians were right. It does reduce some of the more difficult aspects of menstruation and the menopause, especially the number of hot flushes ('flashes' in America), mood swings and night sweats. Black cohosh seems not to have the side effects that conventional hormone replacement therapy sometimes causes. Its own side effects may include nausea, vomiting, uterine contractions and bradycardia. It looks as if its importance is set to increase dramatically.

Black cohosh is also sometimes called black snakeroot, bugbane, bugwort, richweed, squawroot and rattle root, and European blackberry. However, not all plants called snakeroot in folk medicine are synonymous with black cohosh and the innocent plant to which the name blackberry is more usually applied is not even in the same family. The genus *Cimicifuga* is a member of the *Ranunculaceae* or buttercup family and, like most members of that family, including aconite, needs handling with great care. The genus contains eighteen species, six Americans, one European and the rest Asian. The rank-smelling European species was used to discourage bed bugs, hence the name bugbane. Black cohosh is not remotely related to blue cohosh (*Caulophyllum thalictroides*), which is very dangerous.

For a woman able to bear children, whole new floras awaited, according to whether she desired or feared the conception of a child. If she was expected to produce children, she was fed on carrots (*Daucus carota*). In sixteenth-century Europe these were thought to be aphrodisiac, perhaps because the large orange carrot varieties we eat today had then only just evolved and occasioned ribaldry and excessive hopes. Failing carrots, a woman was sometimes suspended over a steam bath of catnip (*Nepeta cataria*), or, once her period was over, made to drink salted sage juice to stop 'abortments' of any subsequent conception. More dangerously, she was given savin (*Juniperus sabina*), though if the dose was too strong it caused abortion instead. In America, she was given sassafras to encourage fertility. In tropical regions, various *Physalis* species are still used for fertility. *P. angulata* is the most widely found. Common names amongst the genus are legion: mullaca, Chinese lanterns, cape gooseberry, winter cherry, tojuá-de-capote, polopa, wapotok. Chinese lanterns is most commonly used for the splendid garden perennial *P. alkekengi*. All are packed with exotic

Actaea racemosa, here pictured in John Torrey's *A Flora of the State of New York* (1843), is a deliciously scented garden plant. The roots were used by native Americans for snakebites, as a mild relaxant and for 'women's concerns'; Western medicine uses it to reduce the discomfort of the menopause.

In a mid-sixteenth-century account of Aztec crafts, written and illustrated by Bernardino de Sahagún, an Aztec midwife rubs a woman's back with leaves, presumably ones held to soothe the pains of labour. Every culture has a huge flora connected with conception and childbirth.

phytochemicals, though few have as yet had much serious investigation. Mullaca (*P. angulata*) is an annual herb indigenous to many parts of the tropics, from central Africa, tropical Asia and the Amazon basin and in cold climates an easy greenhouse annual. Solomon islanders use it to promote fertility. The plant is used in the West Indies and Jamaica to prevent abortion or miscarriages. Rainforest Amazonians also enjoy its sweet if bland fruit. Some of them make a diuretic tea of the leaves. More excitingly, Colombian tribes use it as a narcotic. Brazilian tribes use it for female disorders and jaundice. Its fruit is often used to decorate desserts in Western restaurants.

When pregnancy was not the desired outcome of sexual pleasure, if the woman had a helpful mate, he took a drink made of honeysuckle (*Lonicera* spp.) that was supposed to dry up his semen. Alternatively, just before sex, a bunch of mint (*Mentha* spp.) was pushed into her vagina. A few pieces from the fronds of the female fern (*Athyrium filix-femina*) or a tuft of French honeysuckle (*Hedysarum coronarium*) were thought to do as well. Less mechanical was the drinking of a tisane of the common corn-flag (*Gladiolus communis* subsp. *byzantinus*), which a woman took if she lived in southern Europe. Further north, she was fed the juice from roots of the poisonous male fern (*Dryopteris filix-mas*). Alternatively, she took poplar bark stewed with the kidney of a mule.

If perchance a woman became pregnant, there was a vast and dangerous flora that would abort the foetus and restart her periods satisfactorily. In Europe, the extremely poisonous dog's mercury (*Mercurialis annua*), the wild carrot (*Daucus carota*) and stone parsley (*Sison amomum*) would do the business. Gerard recommended a pessary made of the foul-smelling leaves and roots of the gladdon (*Iris foetida*), which was also used to induce periods generally. Best was wild cyclamen or sow bread, which intentionally pregnant women were not even allowed near lest the child be lost. Curiously, like the carrot, it was also believed to be a wild aphrodisiac. Many aphrodisiacs, even anaphrodisiacs, have been used to procure abortions. The ancient nutmeg, once used to spice the night, became, at some stage in the nineteenth century, used to induce menstruation and procure an abortion. Women poisoned themselves by eating large quantities of it well into the 1950s. Toxic juice from both 'male' and 'female' ferns could stop a pregnancy in its tracks. Even the common foxglove (*Digitalis purpurea*) was sometimes brought in to promote abortion, perhaps stopping the foetal heart before the mother's. Even in the late nineteenth century, it was thought to 'excite strong regular, and intermittent uterine contractions', which were supposed to make it an abortifacient and an anaphrodisiac in women. As the dosages were near fatal, this is not a surprise.

Even the apparently innocent flavourer of cakes and ales, the frilly-leaved and golden-flowered tansy (*Tanacetum vulgare*), had a darker use from medieval times. A strong tincture was used to procure an abortion, restore the menses and cure abscesses of the labia. Various artemisias were popular too. Birthwort (*Aristolochia clematitis*), a member of a genus that still poisons the unwary, especially in some batches of Chinese medicines, was used as a pessary. Used thus, it expelled living and dead foetuses, and afterbirths. Calamints and the chaste tree (*Vitex agnus-castus*), especially when combined in a pennyroyal pessary, did the same. Dittany (*Dictamnus albus*) was also used. The list goes on and on: *Myrrhis odorata* and *Oploponax* in the *Araliaceae* family; pomegranate (*Punica granatum*); rue (*Ruta graveolens*); the squirting cucumber (*Ecballium elaterium*); even the stocks (*Matthiola* spp.) or wallflowers (*Erysimum cheiri*) from the flower borders. Juniper (*Juniperus communis*) was an especially popular abortifacient in the eighteenth century, and contributed much to the vogue for gin, of which it was the main flavouring.

In Europe the most popular abortifacient of all was a plant that is now a rampant inhabitant of the sentimentalist's herb garden: pennyroyal (*Mentha pulegium*). It is too strongly flavoured for the kitchen, but pigs are supposed to enjoy its taste. The common name of pudding grass refers to pigs' food – perhaps pigs have a mechanism for overcoming the plant's dangerous hepatotoxins. In Europe it has been the most important plant to procure abortions since its mention by Dioscorides. All other ancient writers from Hippocrates, Pliny the Elder and Galen, to Macer, Avicenna and Matthiolus describe its use. Macer's herbal (nothing to do with the Macer of ancient times but probably actually written in the early eleventh century by Odo, Bishop of Meung) says: 'Vis prima. If a woman that is with child drinke often this herbe, she shal a'werpe [abort] her childe, but nathetles this herb if groundyn smale and drunkyn in mulsa wine put out a warpel [embryo] of the wombe. Puliole [an old name for Pennyroyal] dronken in leuke-wyn wot bringe oute women flourz

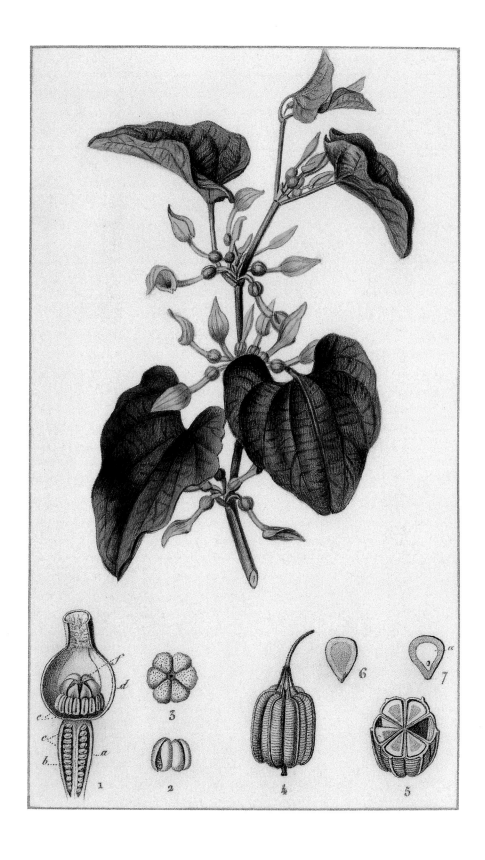

[menstrual products] and in thes same wyse delyuere women of their after-burden [afterbirth].' (He also mentions southernwood, wormwood, betony, chamomile, fennel, lily, chervil, mustard, marjoram, thyme, peony, cinnamon and spikenard as menstrual stimulators.) Not all old herbals mention its properties. In the seventeenth century, Gerard notes that it 'It bringeth forth the monthly termes and bringeth forth the secondine, the dead childe and the unnatural birth.' He says nothing about terminating an ordinary pregnancy. Knowledge of it may have gone underground. In modern times, Mrs Grieve's *A Modern Herbal* (1931) makes no mention of abortion, though she describes pennyroyal as an 'old fashioned remedy for colds and menstrual derangements', which may be a covert signal. In contemporary America, pennyroyal tea is still taken to induce abortion, though the dosage needed to do the deed is so high that it sometimes kills the mother too. Incidentally, again common names cause confusion. What is called American pennyroyal is *Hedeoma pulegioides*, and not the pennyroyal taken to America by European colonists, *Mentha pulegium*; American pennyroyal has different properties.

Although so many doctors and apothecaries have been men, occasional female voices mentioned abortifacients. The nun, composer, doctor-apothecary and beatified Hildegard of Bingen (1098–1179) did not flinch from enumerating seven abortifacients in her *De simplicis medicinae*. These included tansy, oleaster and what she called 'nasturtium'. This was *Nasturtium officinale*, the common watercress. What we now think of as nasturtiums, being species of *Tropaeolum*, were not yet introduced from America. Another twelfth-century woman physician and medical writer, Dr Trotula of Salerno, recommends artemisia wine as an abortifacient or, if that fails, a potent and exceedingly dangerous drink combining hemlock, artemisia, myrrh and sage. Beware of vermouth.

A number of plants contain substances that disrupt or desynchronize pre-ovulatory and pre-implantation events. Some of these substances are isoflavones or uterine contractors. Others are oestrogenic steroids that act on the hypothalamus and pituitary gland. Yet others are prostaglandins. Given that the network of hormones acting on uterus and cervix are carefully balanced, it is not surprising that some plants can aid fertilization at one dosage and cause abortions at others. Of these sorts of Janus plants, savin and sassafras are the most important. Savin (*Juniperus sabina*) was an abortifacient from at least Roman times. Both as we have seen were used to help conception. The handsome sassafras tree of North America (several species of *Sassafras*) was one of the plants joyfully received by Nicolás Monardes in Seville in the sixteenth century. As with so many American plants thought to have some medicinal effect, there was much confusion about what sassafras would actually do. In 1586, Francis Drake, on landing at Roanoke, Virginia, heard tales of colonists surviving on soup

Aristolochia clematitis, the so-called birthwort, by Pierre-Jean-François Turpin (1775–1840). Some sources claim that its name refers to the 'foetus shaped' flowers, and not to its abortifacient properties. Some herbalists claim it heals cancers, though others suggest it at dosages that would cause swift renal failure. Aristolochic acid is extremely toxic.

made from sassafras. He returned to England with the first British shipment of this plant. A few years later, by 1602, Bartholomew Gosnold (who named Cape Cod and Martha's Vineyard) was shipping sassafras regularly to England, and by 1607 it was being sold in every English coffee house – then dens of iniquity – and even on the street. Probably the main reason for this commercial success was that it was supposed, like sarsaparilla, to be a cure for syphilis and gonorrhoea. It was also thought to relieve some sorts of menstrual pain, and pain following parturition. And it was used to cause abortions. In America over the succeeding centuries it was gradually demoted to being a flavouring of root beer. In the 1960s, some of sassafras's constituents were shown to be carcinogenic; it is now hardly used.

As Shakespeare's man, or woman, slips into and through the fifth and sixth stages of man, plants have been called into help. From hair loss to warts, acne to grave marks, slimness to sagging breasts, concerns about bodily appearance have engrossed humans wherever and whenever they have lived, and correspondingly the flora of vanity is colossal.

Perhaps because they particularly show the passing of years, skin and teeth have had many plant treatments applied to them. Almond oil and olive oil were rubbed in to skin to keep it smooth and supple. Gerard suggested linseed oil and the juice of poisonous black briony (*Tamus communis*) for minor skin blemishes. One seventeenth-century Scottish peeress, a Countess of Cromarty, kept a recipe that began:

> *To make the face faire and the breathe sweette, Take the flowers or blossomes of Rosmary, boille them in whytt wynne, then straine out yo blossomes and wasche well youre face, mouthe, teethe and gomes threw the morninge, noune, & Eveninge and let th[a]t drie againe of it selfe. This will make a fayre whytt skynne and make a sweet breathe and will scoure the teethe & gomes greatly & p[re]serve the[m] from the cankar.*

The Cromarty ladies were also worried about their weight, and made themselves a drink from American sarsaparilla, coriander, fennel and aniseed, which they called a 'Diet Drinck'. As it also contained an ounce of antimony for every two gallons made, it must have been exceedingly poisonous. They may well have got thinner.

For wrinkles Gerard suggested walnut oil or juice from the mastick tree (*Lentiscus* spp.). Juice from daffodil bulbs, or the much rarer crinum lily recently brought in from South Africa, was especially good for the wrinkles on ankles. White briony (*Bryonia dioica*), especially when combined with fenugreek, also did for wrinkles, freckles, as well as spots, moles and other imperfections.

A 'new' plant – though in fact it has a long history of use – for which the modern cosmetic industry has high hopes is the African shea tree. Its seed can be ground to make shea butter, which, like the almond or walnut oil used in the sixteenth century, is a delicious and fatty substance said to have remarkable moisturizing properties; it is also supposed to remove wrinkles. The first European to observe the uses of *Vitellaria paradoxa* (syn. *Butyrospermum parkii*) was the explorer Mungo Park. In his *Travels into the Interior of Africa*, published in Edinburgh in 1797, he describes travelling up the Gambia River. He notes that the natives

Mungo Park (1771–1806) was a Scots surgeon for the East India Company, and much interested in medical botany. He made two expeditions to West Africa in 1795 and 1805, and died trying to find the source of the Niger.

… supply the inhabitants of the maritime districts with native iron, sweet-smelling gums, frankincense, and a commodity called shea-toulou, which, literally translated, signifies tree-butter. This commodity is extracted from the kernel of a nut by boiling the nut in water … The extract has the consistency and appearance of butter; and is an admirable substitute for it. It is an important staple in the food of the natives and therefore the demand for it is great … The butter produced from it, besides the advantage of its keeping the whole year without salt, is whiter, firmer, and, to my palate, of a richer flavour than the best butter I ever tasted made from cow's milk. The growth and preparation of this commodity seem to be among the first object of African industry in Bambarra and the neighbouring states; and it constitutes a main article of their inland commerce.

African women were already using it on their skin to give it a magnificent gloss. As today's market for natural and organic cosmetic products gathers pace, the growth potential for shea butter may be huge. It can already be found in anti-wrinkle creams, lip balm, stretch-mark ointments and soothing nappy rash lotions. It is being promoted as a new West African cash crop, and its cultivation is being much extended.

Warts, the skin growths caused by papilloma viruses, have also caused distaste and embarrassment. Even though Oliver Cromwell was happy to have his portrait show his, most humans worldwide try to get rid of them. To this end Fijians use a number of plants from the euphorbia and nettle families which exude corrosive latex when the leaves or stems are broken. The juices very often contain enzymes which break down animal proteins and therefore might make the virus's position untenable. Euphorbia sap was used in sixteenth-century Europe, as was that of plants such as greater celandine (*Chelidonium majus*). One recipe suggested: 'Take an sticke of greene oke woode laie th[a]t in the fier to burne and keepe the water therof w[hi]ch wil ishew out at the ends and therw[i]t[h] anoint the wartes. Use and follow th[a]t well twisse or thrisse a daie rubbing the warts well therw[i]t[h] and th[a]t will take them cleane awaie for euer as cleare as if youe never haid had any.' The savin tree (*Juniperus sabina*), if not being used for women's problems, was used against venereal warts so visible on men who had consorted with 'uncleane women'. Tormentil (*Potentilla erecta*) in strong solution was used against ordinary ones, as well as rheumy eyes and piles.

The Sutherland Sisters sold to a largely female market, but hair can be a matter of anguish for men too and men had a remarkably large flora devoted to stopping them going bald. Even by the seventeenth century, nearly twenty plants were used. The ash of various reeds was mixed with vinegar and applied to the scalp – perhaps it was believed that a bed of reeds looks like a good head of hair. The sticky leaves of *Cistus ladanifer* were plastered on scalps. The juice squeezed from the leaves of the southern nettle tree (*Celtis australis*) was also supposed to do the deed. An early-eighteenth-century manuscript suggests to barbers: 'beat linseeds very well: and mix them with sallet oyl [salad oil] and when you have well mixed them anoint the head therwith and in three or four times using it it will help you.' At least these plant substances were easily available. Another recipe suggested: 'for beldiness in the head Anoint with the oil of Luzeards [lizards] and the beldness will grow one again'. Presumably the regrowth of a lizard's shed tail suggested that lizard extract would regrow hair.

If the wheel of fortune keeps spinning, whatever worries an individual has about wrinkled skin or greying hair give way to the wheel's slowing spin, and thoughts of what will happen when its movement stops. There are two great fantasies in human life. Colossal wealth is one. Life everlasting is the other. Often in the past the rare humans who had the first naturally believed that they should have the other. The desire to avoid death was as powerful four millennia ago as it is today.

A Chinese emperor may have been the first to look for such a thing as an elixir of life: Ch'in Shi Huang Ti, who ordered the building of the Great Wall. As the boy-king Cheng,

The first Chinese emperor, Ch'in Shi Huang Ti, believed too much in his magicians, and could not accept that he, like all his subjects, was mortal. He made no clear plans for the succession, and his new empire fell apart when he died in 210 BC.

he came to the throne of one of the states that made up what we now call China in 246 BC. Warlike, energetic and very clever, he conquered the adjoining states and united them into a single entity. In 221 BC he created the Qin dynasty and was the first in Chinese history to use the term 'emperor'. He built massive palaces and enhanced the ceremony of the court to further his own glory and overawe the people. He replaced the old feudal system with a centralized bureaucracy. Old royal or aristocratic families were banished. Scholars were killed. Books, except technical ones on law, agriculture and medicine, were burned. Not surprisingly, he made many enemies, and became so fearful for his life that he lived in complete secrecy. He also became obsessed with obtaining immortality and had his empire ransacked for plants that might form an elixir of eternal life. Nothing helped. He died in 210 BC in Shandong Province, far from his capital of Changan. He probably poisoned himself, trying something he'd found during one of his long quests to find the impossible secret. His empire rapidly descended into chaos.

Many subsequent emperors tried to find what had eluded the first. Lesser Chinese looked too. They placed great hopes in a plant that has been intriguing Westerners since they first heard of it in the seventeenth century and is still widely popular. There are two species of ginseng. The Asian ginseng is *Panax ginseng*, and it is one of the five astral remedies in Chinese medicine. The very similar American ginseng is *P. quinquefolius*. A third plant, called Siberian ginseng (*Eleutherococcus senticosus*), is only very distantly related. The Asian ginseng has been in use in China for well over two and a half thousand years. During that time, it has become so widely used by so many people, and taken by each person for such considerable lengths of time, that even in China wild material is rare and expensive. Many other plants have been used as substitutes. It is supposed to increase longevity.

The North American species was in use by native Americans at least from the sixteenth century. They used it mostly as a constituent of love potions. Under that guise, it reached a hopeful Europe in 1704. The first scientific paper on its more general use was published by Michel Sarrasin, who had brought roots of American ginseng to Paris. A paper he presented on this topic was published in the proceedings of the French Academy for 1714. In that same year, a missionary to China, Father Jartoux, published an article on Asian ginseng in a London scientific journal. Though a traveller called Samedo Alvaro recounted stories to Europeans about the Chinese healing root called jin-chen, or ginseng, in 1642, by the time of Jartoux's paper, Chinese apothecaries were desperately concerned about the dwindling supplies of their own life-lengthening species. Enterprising merchants pricked up their ears.

A surprising trade developed. In 1718, the first shipment of American ginseng arrived in the Chinese port of Canton. The Chinese took to it gradually. Trade increased, and by 1773, a shipment of 55 tons was sailing from Boston aboard the *Hingham*. In China it earned nearly $3 a pound – a good profit. Soon all suitable parts of North America were filled with native and European ''seng diggers' hunting for good roots. Even the American explorer and pioneer

Daniel Boone (1734–1820) got involved, selling fifteen tons of American ginseng to Philadelphia merchants in 1788. Naturally, wild supplies of the American plant soon gave out. Something had to be done. By 1885 George Stanton had founded a 150-acre ginseng farm in New York. Others followed suit, and ginseng is now extensively cultivated.

All parts of both species contain triterpenoid saponins called ginsenosides. These are thought to be responsible for whatever pharmacological activity ginseng might possess. It is often supposed to improve concentration, promote a better grasp of abstract concepts, and increase visual and motor coordination. It is supposed to reduce fatigue as well. All these properties ally it with the much more feared coca and qat. However, unlike them, it is also used in an attempt to improve well-being and, most desired of all, longevity. But then it is supposed to cure depression and anxiety, general debility, male impotence, spermatorrhoea, stubborn fevers and persistent coughs; to act against nausea in pregnancy, dyspepsia and chronic malaria, and to boost the immune system. Such an extraordinary range of effects suggests that the placebo effect is working very strongly. Certainly, none of these medicinal effects has any scientific basis at present. It is also sometimes claimed to be a natural alternative to oestrogen that might help women going through menopause. In fact ginseng has no oestrogen-like properties. The only thing it has so far been shown to do is reduce blood glucose levels in people with certain types of diabetes. It should not be taken in combination with insulin or oral anti-diabetes agents. Its overuse, for people lucky enough to find a product with a real ginseng content, can cause insomnia, and more rarely diarrhoea, vaginal bleeding, headaches and even schizophrenia. It also interacts with the drug warfarin, reducing its anticoagulant effect. Since 1994 ginseng's annual sales in America have exceeded $200 million; 220,589 pounds of ginseng were harvested from the wild in 1992. Ageing is big business.

If the emperors of China, men of huge wealth and power, were unable to cheat death, there was little chance for anyone else. In China, as in Egypt, and later Arabia and medieval Europe, one class of person continued to proclaim that the elixir of life must exist, and had been, or was just about to be, found: alchemists. While most alchemists settled for the pursuit of limitless supplies of alchemical gold, others wanted the great elixir to stop the progress of the seven ages of man, and hold Death for ever at bay. Some experimented first with plants, incinerating mandrake roots and sealing their ashes into glass flasks. The flasks were incubated in beds of decaying horse manure, and the mandrakes regrew. They tried it with animals. Giambattista della Porta proposed in his *Magia Naturalis* to show 'how living Creatures of divers kinds, may be mingled and coupled together, and that from them, new, and yet profitable kinds of living Creatures may be Generated [and] ... how to generate pretty little dogs to play with'. This was nothing. Paracelsus, of course, claimed to have gone much further. In *De Natura Rerum* of 1572, he wrote:

A nineteenth-century wood engraving shows a nonchalant Daniel Boone. This intrepid pioneer, hunter and explorer was also the owner, and loser, of several extensive farms. He famously once remarked: 'I have never been lost, but I will admit to being confused for several weeks.'

Let the semen of a man putrefy by itself in a sealed cucurbite with the highest putrefaction of venter equinus for forty days, or until it begins at last to live, move, and be agitated, which can easily be seen. At this time it will be in some degree like a human being, but, nevertheless, transparent and without a body. If now, after this, it be every day nourished and fed cautiously with the arcanum of human blood, and kept for forty weeks in the perpetual and equal heat of venter equinus, it becomes thencefold a true living infant, having all the members of a child that is born from a woman, but much smaller. This we call a homunculus; and it should be afterwards educated with the greatest care and zeal, until it grows up and starts to display intelligence.

If these beings could be kept alive, they grew into 'giants, pygmies, and other marvellous people, who get great victories over their enemies, and know all secrets and hidden matters'. If they died, they were pickled and put into bottles; these are shown as a prop in many seventeenth-century illustrations of alchemists' laboratories.

Paracelsus, like every other human, died. No magical mandrake, no ginseng, no plant of any sort has ever yielded a true, or even a partial, elixir. One alchemist did, indeed, claim to have been given a few drops of the amazing substance. The book now exists as seventeenth-century copies of something supposedly written in the twelfth century by a self-deluded or deliberately prankish alchemist. The book is *De Vita Propaganda* (The Art of Prolonging Life). At its close, the writer proclaims:

I, Artephius, having learnt all the art in the book of Hermes, was once as others, envious, but having now lived one thousand years or thereabouts (which thousand years have already passed over me since my nativity, by ... the use of this admirable Quintessence) ... I have seen, through this long space of time, that men have been unable to perfect the same magistry on account of the obscurity of the words of the philosophers. Moved by pity and good conscience, I have resolved, in these my last days, to publish in all sincerity and truly, so that men may have nothing more to desire concerning this work. I except one thing only ... Nevertheless, this likewise may be learned from this book, provided one be not stiff-necked and have a little experience.

The roots of ginseng (*Panax ginseng*) take several seasons to grow large enough for harvest, and are now rare in the wild. Commercially cultivated material accounts for the bulk of world supply, and is expanding. In spite of that, there is, as yet, no clear evidence that the plant has much effect on human health.

THE MIND

*O*ne December evening ... I arrived in a remote quarter in the *middle of Paris, a kind of solitary oasis which the river encircles in its arms on both sides as though to defend it against the encroachments of civilisation. It was in an old house on the île St. Louis, the Pimodan hotel built by Lauzun, where the strange club which I had recently joined held its monthly séance. I was attending for the first time.*

Though it was scarcely six o'clock, the night was black. A fog, made thicker still by the nearness of the Seine, blurred all the shapes under its quilting ... The pavement, inundated with rain, glistened under the street lamp as water reflects an image; a sharp dry wind carrying particles of sleet whipped into the face ... None of winter's rude poetry was wanting that night.

Théophile Gautier (1811–72), novelist, poet and art critic, was writing for the *Revue des Deux Mondes* on 1 February 1846. The club was the Club des Haschichins, founded in 1835. By the time Gautier became a member, it was housed in stylishly decayed grandeur at the apartment of Dr Jean-Jacques Moreau, who initiated the science of psycho-pharmacology in France, and used cannabis to treat the insane and depressed. Gautier was eventually admitted to an apartment, where the other members were waiting. Moreau lived in a style that deeply appealed to the romantic Gautier. The frames of the paintings had been sold for the value of their gold leaf, tapestries were grand but frayed and faded, crockery chipped and cracked, though Oriental. Gautier goes on:

In a work of the Boucicaut Master (*fl.* 1390–1430), Marco Polo, suitably acccompanied by camels and an elephant, enters Hormuz, where the power of plants, especially *Cannabis sativa*, was well known, and even played an important role in local politics.

The doctor stood by the side of a buffet on which lay a platter filled with small Japanese saucers. He spooned a morsel of paste or greenish jam about as large as the thumb from a crystal vase, and placed it next to the silver spoon on each saucer.

The doctor's face radiated enthusiasm; his eyes glittered, his purple cheeks were aglow, the veins in his temples stood out strongly, and he breathed heavily through dilated nostrils. 'This will be deducted from your share in Paradise,' he said as he handed me my portion. After each had eaten his due, coffee was served in the Arab manner ...

During the subsequent, and determinedly exotic, banquet, he notices:

My neighbours began to appear somewhat strange. Their pupils became big as a screech owl's; their noses stretched into elongated proboscises; their mouths expanded like bell bottoms. Faces were shaded in supernatural light. One among them, a pale countenance in a black beard, laughed aloud at an invisible spectacle; another made incredible efforts to raise his glass to his lips and the resulting contortions aroused deafening hoots from his companions; a man, shaken with nervous convulsions, turned his thumbs with remarkable agility; another, fallen against the back of his chair, his eyes unseeing and his arms inert, let himself drift voluptuously in the bottomless sea of nothingness.

My elbows on the table, I considered all this with clarity and a vestige of reason which came and went by intervals, like the light of a lantern about to flicker and die. A deadening warmth pervaded my limbs, and dementia, like a wave which breaks foaming onto a rock, then withdraws to break again, invaded and left my brain, finally enveloping it altogether. That strange visitor, hallucination, had come to dwell within me.

Gautier, the arch-romantic, went on to have, he claimed, an unnerving and frightening 'trip'. Part of it may have been due to his expectations. Indeed, the individual response to cannabis, and to many other plants in this book, is determined very much by the culture of the taker. This effect has been known since the sixteenth century or so, when observers noticed that smoking tobacco allowed American Indians to reach intense trance-like states, whereas in the Old World smokers rarely reached much more than a mild, if pleasant, state of relaxation. Even within the same general culture, 'mind' drugs work in different ways on different individuals. Some people find that cannabis, whether smoked or eaten, brings on a pleasing dreamy state. Sometimes they recall long-forgotten events. Sometimes thoughts loose their linear connections. Sometimes the perception of time, and sometimes of space, is altered. Dr Moreau must have given large helpings of the 'green mustard' to induce Gautier's extreme visual and auditory hallucinations. In any case, the romantic writer was determined to have the sort of experience that would make good copy for an article. He

Théophile Gautier, all dressed up and ready for anything, was shaken by his encounter with Dr Moreau's 'green mustard' at the Club des Haschichins in 1845. Moreau was also experimenting with the use of *Cannabis sativa* on those of his patients who had severe psychological problems.

remained observant enough, though, to note the slighter reactions of the other club members. Most felt euphoria, excitement and inner happiness. Many laughed and laughed. And he barely mentions any feeling depression as the 'high' subsided. Dr Moreau's patients would have had an interesting time of it.

The hashish Gautier had taken was a conserve made of the flowering tops of highly potent forms of hemp (*Cannabis sativa*). In taking this, he was indulging in something known to the oldest civilizations of the Indus. Nowadays this extraordinary plant is entirely global. The tall stems give a fibre that can make fabrics as coarse as sacking, or almost as fine as linen. The fibres, until recent times, made almost all the ropes that made commerce, building and travel possible. The seeds are edible and nutritious, and can be pressed to give an excellent oil. They store well and are easily transported. The leaves have been used medicinally to help an astonishing range of ailments, by the Scythian nomads of the ancient world, by practitioners of Chinese and Ayurvedic medicine, by the modern Khoi and San tribe wanderers of the Karoo and Kalahari deserts. Because substances in the leaf also affect the mind, the plant has become more deeply involved with the quirks of human morality than almost any other, except perhaps the poppy. Its illegality in many parts of the world, the first attempt at its suppression dating from 1379, and in modern America from 1915 onwards, underpins a large percentage of the 'black' economy, of crime and suffering. That was not always so.

The crop is so ancient that its place of origin is unknown, though it was first domesticated somewhere in Central Asia. Man's partnership with the plant is ancient: he has used it for ten thousand years, probably as one of the very first of his crops. The earliest known hemp fabric has been found in Chinese excavations of sites dating from 4000 BC. Hemp rope and thread from Turkestan have been dated to 3000 BC. The crop had developed a medicinal role in China by 2800 BC. In the *Pen Ts'ao* (Pharmacopoeia) Shen Nung describes it as a 'superior' herb. The text notes that ma-fen (hemp fruit) would 'if taken to excess ... produce hallucinations [literally, seeing devils]'. In lesser doses, it was useful for malaria, beri-beri, constipation, rheumatic pains, absent-mindedness and female disorders. Hoa-Glio, another ancient Chinese herbalist, suggested it, mixed with wine, as a painkiller.

The plant's main distributors were almost certainly the wild Scythian tribes. Originating in Central Asia, these tribes migrated north and west, taking their marvellous and hugely useful herb with them, and reached Europe from about 1500 BC. About 500 BC the Greek writer Herodotus reported that:

... they make a booth by fixing in the ground three sticks inclined toward one another, and stretching around them woollen pelts which they arrange so as to fit as close as possible: inside the booth a dish is placed upon the ground into which they put a number of red hot stones and then add some Hemp seed ... immediately it smokes and gives out such a vapour as no Grecian vapour bath can exceed; the Scyths, delighted, shout for joy.

A plate from *A Children's Picture Book* (1796), showing *Cannabis sativa* plants. The upper parts of the plant are the richest in cannabinol.

Archaeologists have recently excavated frozen Scythian tombs in Central Asia, dated between 500 and 300 BC, which, as Herodotus suggested, contained tripods and pelts, braziers and charcoal with the remains of cannabis leaves and fruit to help the buried cope with the stresses of the afterlife.

The plant migrated southwards into India. There it was such a success that legends say that the gods themselves sent it as a present to man, so that he might attain delight, desire and courage. Whenever nectar or amrita dropped down from heaven, cannabis sprouted from the fall. Consecrated to Shiva, it flavoured the goddess Indra's favourite drink. Since it was therefore a plant of the gods, its use was believed to bestow supernatural powers on its users. It was used in Hindu rites at least by 600 BC and probably long before. In the Vedas, it is a plant able to give man anything from good health and long life to visions of the gods. It quickened the mind, improved judgment, lowered fevers, induced sleep, cured dysentery. The medical work *Sushruta* claimed that it cured leprosy.

By AD 1600, it had collected even more properties, being by then considered an anti-phlegmatic, a good digestive and a stimulant to the appetite, and even to give the taker a more beautiful voice. If all that wasn't enough for any one herb, it could also be used for the control of dandruff and headaches, mania, insomnia, venereal disease, whooping cough, earaches and tuberculosis. If the sufferer's problems were spiritual, he took a preparation called bhang, which was so sacred that it banished evil, brought luck and cleansed man of sin. Sacred oaths were sealed over hemp. The Hindu god Shiva commanded that the word *bhangi* must be chanted repeatedly during the sowing, weeding and harvesting of the holy plant. *Bhang* is a paste of dried leaves or flowering shoots of cannabis, spices, flour and sugar. Like the similar sweetmeat *maajun*, it is also found right across North Africa, though it is sometimes laced with poppy seeds, flowers of the thorn apple (*Datura* spp.), and even some nux vomica. Théophile Gautier's experience may have been, in part, due to these additions. It was sometimes also used as a painkiller. A British soldier, held captive by rebels in Mysore, wrote in 1781: 'Our ill-favoured guard brought in a dose of majum each, and obliged us to eat it ... a little after sunset the surgeon came, and with him 30 or 40 Caffres, who seized us, and held us fast till the operation [circumcision] was performed.'

In spite of its colossal repertoire further east, in Greece and Rome cannabis seems hardly to have been much more than a fibre plant and a medicine. However, Democritus suggests that it was occasionally drunk with wine and myrrh to produce visionary states, and Galen, about AD 200, wrote that it was customary to give cannabis to guests to promote hilarity and enjoyment. Perhaps with their extensive vineyards and excellent wines, the Greeks and Romans had little need for other stimulants.

It was in the post-Muhammadan Middle East that the first prohibitions took place. Muhammad, proscribing alcohol, permitted the use of cannabis. It was in wide use, certainly from the eighth century or so. By then the stories that make up *The Thousand and One Nights* (or *The Arabian Nights*) were beginning to be told. Hashish played a large part in some of them,

During the craze for all things Oriental in late-nineteenth-century Europe, Scheherazade fascinated artists. She is shown here in a painting by Edouard Frédéric Wilhelm Richter (1844–1916).

particularly in its tales of modest men becoming extremely immodest under its spell. Scheherazade begins the 'Tale of the Hashish Eater' to her dangerous owner, Sultan Shahriyar:

> *Then said she, 'Know that I mean to pass this night with thee, that I may tell thee what talk I have heard and console thee with stories of many passion-distraughts whom love hath made sick.' 'Nay,' quoth he, 'Rather tell me a tale that will gladden my heart and [make] my cares depart.' 'With joy and good will,' answered she; then she took seat by his side (and that poniard under her dress) and began to say: – Know thou that the pleasantest thing my ears ever heard was …*

She goes on to tell a funny story about a man's outrageous behaviour under the influence of the plant. Though the licentiousness of hashish takers may have been one reason for the plant's prohibition, it is likely that there were several causes. Certainly, one was the increasing power of theologians, and their desire for a stranglehold on the availability of access to mystical states. Another may have been an attempt to stifle the residual power of

The Old Man of the Mountains, al-Hasan ibh-al-Sabbah tells the young Assassins what
needs doing, as an assistant offers the 'certain potion' to a rather doubtful-looking boy. This
illustration of Marco Polo's travels is by the Boucicaut Master (*fl.* 1390–1430).

an early terrorist group known as the Assassins. Their ideas developed out of the Ismaili
section of the Shiite strand of Islam. The word itself is supposed to be a corruption of the
Arabic word for hashish user: hashishiyya. Modern Ismailites prefer to believe that it is
derived from the word for followers of Hasan.

The first Assassin was a young aristocrat called al-Hasan ibn-al-Sabbah (d. 1124). He was
probably a Persian from Tus. After spending a year and a half in Egypt, he became a Fatimid
missionary. Once home, and with a band of followers, he took by force the almost
impregnable mountain fortress of Alamut, in the Elburz Mountains. His 'eagle's nest' became
a central stronghold from which his henchmen gradually expanded their territory. Being a
small group, disciplined and secretive, they made use of assassination rather than more open
forms of warfare. Their beliefs were radical. They aimed to emancipate the initiate from the
toils of empty doctrine, suggested that all prophets were superfluous to individual revelation,
and encouraged members to believe nothing and to dare all. Assassinism was exciting and
anti-authoritarian, and attracted many young adherents. The Venetian traveller Marco Polo

visited Alamut in 1271 or 1272, some time after its fall. Describing the magnificent garden surrounding the pavilions and palaces built by the grand master, he writes:

Now no man was allowed to enter the Garden save those whom he intended to be his ASHISHIN. There was a fortress at the entrance to the Garden, strong enough to resist all the world, and there was no other way to get in. He kept at his Court a number of the youths of the country, from twelve to twenty years of age, such as had a taste for soldiering ... Then he would introduce them into his Garden, some four, or six, or ten at a time, having first made them drink a certain potion which cast them into a deep sleep, and then causing them to be lifted and carried in. So when they awoke they found themselves in the Garden.

When therefore they awoke, and found themselves in a place so charming, they deemed that it was Paradise in very truth. And the ladies and damsels dallied with them to their hearts' content ... So when the Old Man would have any prince slain, he would say to such a youth: 'Go thou and slay So and So; and when thou returnest my Angels shall bear thee into Paradise. And shouldst thou die, natheless even so will I send my Angels to carry thee back into Paradise.'

The 'certain potion' contained *Cannabis sativa*.

The assassinations were effective enough to destabilize the world of Islam until the Mongolian Hulagu, who also destroyed the caliphate, seized the Assassins' main fortress in 1256. However, other Assassin strongholds lasted for another few decades, until the Mamluk Sultan Baybars cleared them out too. The Assassins scattered through northern Syria, Persia, Oman, Zanzibar and especially India. The movement still has a large following, adherents being called Thojas or Mowlas, but more often Nizaris. In an attempt at crushing them, in 1378–9, an Arabian emir, Soudon Sheikhouni of Joneima, prohibited cannabis consumption amongst the poor, destroyed all the crops he could find and punished offenders by pulling out their teeth.

In Europe, hemp appears in the Hochdorf Hallstatt D wagon burial site of about 500 BC. It grows perfectly well in northern Europe, and was probably widely grown. Hemp ropes were certainly used throughout the Roman Empire, and an English find is dated to AD 140–80. The plant is wind pollinated, and its pollen survives well in the soil. Pollen counts show that there was a huge upsurge in its cultivation from AD 400 to 1100. However, there is no record of its medicinal or other uses. Nevertheless, and even under a northern sun, anyone harvesting the crop would have been aware of its psychotropic properties. The active substance, cannabinol, is present in the resinous hairs on leaves and stems, and is absorbed, albeit in small quantities, through the skin. There seem to be no early references to its use as an intoxicant, though. During the Crusades, many Europeans would have encountered its widespread use in the East. By the fifteenth century, its use in Europe was apparently widespread enough to alarm the Church. In 1484, Pope Innocent VIII outlawed it, proclaiming that it was an unholy sacrament of the Satanic mass. Plants more relevant to such occasions, such as henbane and mandrake, he left alone.

Whatever Satan may have had to do with it, rope held world trade together. It was

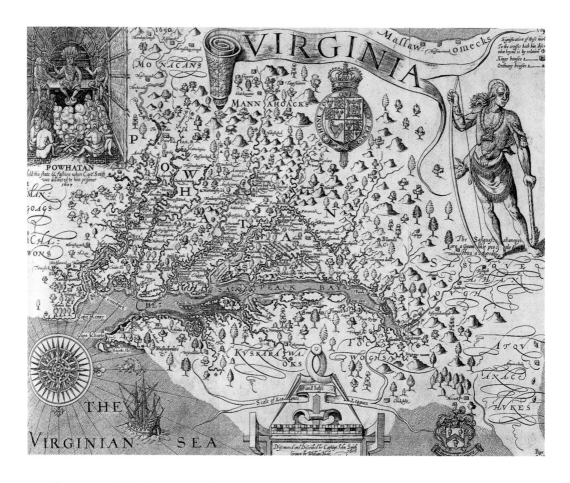

Virginia in 1606. The colony of Jamestown gave its name, if shortened, to the Jimson weed (*Datura stramonium*) after that plant's narcotic foliage, eaten unwittingly, caused the local garrison to go berserk in 1676. The colony was also a centre for hemp growing.

indispensable for ship's rigging, and in every aspect of haulage. It had to be planted wherever trade took place. In 1554, Spanish settlers cultivated it in Peru. In 1563, Elizabeth I of England decreed that landowners with more than 60 acres must grow hemp as one of their crops or be fined £5. In the next year, Philip II of Spain ordered hemp to be grown throughout his new empire, which stretched from what is now Argentina right up to what is now Oregon. British settlers grew it wherever they settled, and in 1619, Jamestown Colony, Virginia, made hemp cultivation mandatory. Similar laws were passed in Massachusetts in 1631 and in Connecticut in 1632. Hemp was even used for barter or to pay taxes. It grew all over New England. In 1637, the General Court at Hartford, Connecticut, ordered all settlers to plant one teaspoon of cannabis seeds. Massachusetts soon followed.

It isn't clear whether or not any of these crops were also devoted to getting high.

Nevertheless, even the most right-thinking Puritans must have realized, amongst the cut sheaves, that there was more languor in the late summer air than the weather warranted. Certainly, books appearing throughout the century praised the medicinal effects of the plant. The *Anatomy of Melancholy* by Robert Burton (1577–1640), published in 1621, suggested its use as a treatment for depression, something to which American settlers were prone. If anyone was truly aware of the psychoactive effects, no one made any fuss about the crop. In the momentous year 1776, patriotic women ran 'spinning bees' to clothe Washington's troops in hempen fabrics. In *Common Sense*, Thomas Paine lists cordage, iron, timber and tar as America's four essential natural resources, writing, 'Hemp flourishes even to rankness, we do not want for cordage.' The first and second drafts of the Declaration of Independence were written on Dutch hemp paper.

On 16 March 1791, Thomas Jefferson wrote in his journal:

The fact well established in the system of agriculture is that the best hemp and the best tobacco grow on the same kind of soil. The former article is of first necessity to the commerce and marine, in other words to the wealth and protection of the country. The latter, never useful and sometimes pernicious, derives its estimation from caprice, and its value from the taxes to which it was formerly exposed ... Hemp employs in its rudest state more labor than tobacco, but being a material for manufactures of various sorts, becomes afterwards the means of support to numbers of people, hence it is to be preferred in a populous country.

He disliked tobacco because 'This plant greatly exhausts the soil. Of course, it requires much manure [but] yielding no nourishment for cattle, there is no return for the manure expended ...' His slaves did the harvesting of the hemp, and probably got the 'high'.

Nine years later, he would have read with interest that Napoleon was so alarmed about the effect that hashish was having on the disaffected troops of his Egyptian campaign that he tried to prohibit its use. His efforts had little effect. Over subsequent decades, the 'mind' effects of hemp were occasionally observed scientifically, but more often gave rise to overheated articles, like Théophile Gautier's, or those of the even more susceptible American Fitz Hugh Ludlow. There were a number of sensational novels in which frightful crimes were carried out under its influence. Meanwhile, pharmacists in Europe and America got on with selling it. Even in chill Edinburgh, Smith Brothers were selling extracts of cannabis to numerous clients by 1857. In 1860, the Ohio State Medical Society thought cannabis excellent for treating as wide a range of maladies as the Ayurvedic doctors of India had treated with it several thousand years before: it was believed to be good for neuralgia, nervous rheumatism, mania, whooping cough, asthma, chronic bronchitis, muscular spasms, epilepsy, infantile convulsions, palsy, uterine haemorrhage, dysmenorrhoea, hysteria, alcohol withdrawal and loss of appetite. Back in Egypt, though, it was becoming a problem, and the then Emir made possession of cannabis a capital offence in 1868. Perhaps this added to its charms. In America, at the Centennial Exposition of 1876 in Philadelphia, there was a much admired Turkish Hashish Exposition, where visitors could 'enhance their fair experience'. It plainly worked, for by 1883 there were hashish smoking parlours in most large cities, and there were supposed to be five hundred in New York alone.

In 1877, the Sultan of Turkey followed his Egyptian underling and made cannabis illegal, though his subjects took little notice. In 1894, the British Indian Hemp Drugs Commission studied the social uses of cannabis and came out firmly against its prohibition. But in the following year the word 'marijuana' was heard amongst Pancho Villa's supporters in Sonora, Mexico. The word was taken up in North America in preference to hashish. During the subsequent Spanish American War, Pancho Villa's army seized the substantial Mexican assets of American press baron William Randolph Hearst. His newspapers swung into action. Both marijuana and Mexicans began to be demonized. However, in 1912, although a huge international conference at the Hague called for the suppression of opium, morphine, cocaine and heroin, it did not consider cannabis dangerous enough to recommend proscribing it. Nevertheless, over the next few years various American states banned it, though some preferred to ban alcohol and leave cannabis legal. The fight against it attracted some interesting people, notably Harry J. Anslinger, who became head of the newly formed Federal Bureau of Narcotics in 1931. He was appointed by his future father-in-law, the banker and Treasury Secretary Andrew Mellon. In 1937, using the curious Machine Gun Transfer Act as a model, he was able to ban hemp without contravening constitutional rights. He claimed that marijuana was the most violence-causing drug in the history of mankind. The objections of the American Medical Association were ignored. He later wrote in his autobiography, *The Murderers*, that for years he illegally supplied Senator Joseph McCarthy with morphine. Even so, when, in 1944, the New York Academy of Medicine reported that marijuana use does not cause violent behaviour, provoke insanity, lead to addiction or promote opiate usage, Anslinger threatened doctors who carried out cannabis research with gaol.

Nowadays, the confusion continues. Cannabis is universally available. It is the badge of the young or alternative. It is essential to a whole part of Western culture. It can be found growing in penthouses in Manhattan, floodlit garages on Scottish islands, backyards in Provence. Yet in some places, possession can lead to lengthy prison sentences. The long-term sufferers of multiple sclerosis plead for it. Our elected representatives uphold its criminality, whilst their own children smoke it cheerfully in the garden shed.

The current moral, rather than medical, confusion over *Cannabis sativa* is not unique. There are many other plants about which we are equally confused. One, which has as yet played hardly any role other than in its native lands, but which may well turn out to be as disruptive as cannabis, waits in the wings. On 4 January 1761, an expedition set off from Copenhagen on a Danish military ship. It was in the pay of Frederick V of Denmark, encouraged by a leading Hebrew scholar of the day, Johann David Michaelis of Göttingen. The expedition was to make a scientific exploration of Egypt, Syria and Arabia, the first of its kind, but it was also expected to shed light on Jewish and Christian myths. The team consisted of Friedrich Christian von Haven (Danish linguist and orientalist), Pehr (or Peter) Forrskal (Swedish botanist), Christian Carl Kramer (Danish physician and zoologist), Georg Baurenfeind (artist, from southern Germany), a Swedish

Kårsten Niebuhr was a man of modest talents, whose astonishing grip on life took him on an extraordinary journey through the Middle East and India. He persisted long after all his companions had been killed by disease, thieves and hardship. He was the first European to describe the effects of taking qat.

ex-soldier named Berggren and a German surveyor called Kårsten Niebuhr. No leader had been nominated, and even by the time they reached the port of Alexandria, almost everyone had quarrelled with everyone else.

The man who concerns us here is Niebuhr (1733–1815). He started life as a poor farmer, studied surveying and, for reasons not now clear, at the age of twenty-seven found himself at sea. He was not psychologically adapted to being a traveller. He was easily frightened. He believed that northern Europe was the true model of civilized values. He reads, at least in the 1792 English translation of his subsequent *Travels through Arabia and Other Countries of the East*, as an unimaginative and rather depressed man. He was, however, a survivor. Every other member of the expedition was dead by 1764, but he went on to India, Persia and Syria. He found the plant in question in the Yemen, where it was sold in every market. The only effect he noted was that it kept him awake. His may not have been absolutely the first mention of qat, though he is almost certainly the first European to have tried it. Barthélemy d'Herbelot de Malainville's *Bibliothèque Orientale* of 1697 mentions what may be a qat drink, 'Cahuat al Catiat', though this may refer to another Yemeni drink, more often called qisher, made from the husks of coffee beans. The confusion is not helped by the fact that Barthelemy had never been to Yemen or East Africa. Nowadays, qat's admirers find that it has properties very similar to those sometimes ascribed to hemp. In modern Yemen, it is believed to relieve colds, fevers and headaches, and dispel depression. It seems to give physical energy and endurance, and enhance mental concentration and stamina. 'Qat widens the mind,' is a common saying. Scholars use it. Businessmen use it. Because it is a cheerful communal activity too, qat is something that 'gathers people'.

Qat (*Catha edulis*) is a small, drought-resistant tree that grows abundantly in suitable areas from southern Arabia through to the high lands of East Africa. It isn't clear where it originated. Most cultivated forms are sterile, which suggests that it has been in cultivation for a very long time. Widely farmed, it is more productive than wild, and fertile, material. It also isn't clear where its narcotic qualities were first discovered, though Ethiopia seems likely.

The walled city of Aden in the Yemen, here shown in a sixteenth-century illustration
depicting the Portuguese assault of 1513, from Gaspar Correia's *Lendas de India*. Qat had
been brought into its markets from the surrounding hills, each one here capped with a castle,
since at least the eleventh century.

Various Yemeni stories account for its arrival. One tale is that qat was brought to Ethiopia
by Alexander the Great, to whom God had given it as a gift for Yemen, and that it is the
sacred laurel of the Greek oracle at Delphi. It was supposed to cure an epidemic of
melancholy. Other stories suggest that both coffee and qat reached Yemen via northern
Arabia in the eleventh century, qat having originated in Turkestan and Afghanistan.
Alternatively, a Yemeni saint of the thirteenth century trying to convert Ethiopia discovered
qat in Harrar. Becoming an enthusiast, he took it home. In fact, it grows wild in both
countries, though it is more abundant in Ethiopia and other parts of mountainous East
Africa. Whether it was taken for pleasure first in Yemen, Somalia or Ethiopia, a sixteenth-
century document notes that qat was expensive, and only used by the rich. Poor poets wrote
of their frustrated desires for it, and it developed a poetry of longing. That suggests that the
first material was gathered, expensively, from the wild.

In ancient times, Muslims chewed qat in lieu of alcohol, the devout often using it to
remain alert for all-night prayer vigils. It was given to thirteenth-century soldiers to reduce

fatigue. Taking it for recreation was probably associated with it becoming a farmed crop, something that may have happened in the southern Red Sea region around the same sort of time. Soon it was chewed as ardently by bored merchants as by mystics. Today, estimates suggest that over 90 per cent of Yemeni men chew qat on a regular basis, and that in nearby countries where it is legal it is equally popular.

Qat contains cathinone, a natural amphetamine. That soon decays to cathine, which is much less potent. The first is a Schedule I drug in America, as are heroin and cocaine. The second is a legal Schedule IV substance. The leaves also contain a small amount of ephedrine, a substance also listed under Schedule I. Qat chewing, if the leaves are fresh, raises alertness, concentration, friendliness, contentment and the free flow of ideas. Some researchers associate it with an increase in aggression and fantasies of personal power. Some men find it aphrodisiac. Their women disagree.

Qat can have negative effects including constipation, haemorrhoids, hernias, paranoia and depression. While Yemenis deny qat's adverse health effects, they freely admit to an insomnia-caused exhaustion that partially disables their ability to work the following day. Niebuhr was correct. In some ways what is more interesting, in that it links qat with the next two herbs in this chapter, is that it inhibits hunger. Enthusiasts for it lose weight and can have low blood sugar.

Though much loved by those who use it, qat has never become a widely traded commodity, because of the curious twist of its biochemistry, by which as soon as the young shoots are picked, the active substance cathinone begins to turn into almost inactive cathine. After one or two days, the leaves are little more psychoactive than cabbage. Even drying the leaves in the hot Yemeni sun does not preserve their properties. There have been only a few attempts at using the herb at long distances from Arabia Felix. In 1917 a London pharmacist imported a batch of dried material from Aden, and set about marketing it. Among the products were tonics called 'catha-cocoa milk', 'chat milk' and another medicine which combined qat with phenolphthalene, described as 'slightly Laxative and tonic'. Qat pills were also produced, and though none can have contained much cathinone, they were on sale for a couple of decades. The plant is an exceptionally rich source of vitamin C, so perhaps purchasers derived at least some benefit for their outlay. Nowadays, consignments of fresh leaves are sometimes flown to homesick Yemenis worldwide. Leaves are also freeze-dried. There must be large areas of the subtropical region of the Americas and of Asia where the plant would flourish, so it may become of more global importance. It is already grown in Australia.

Qat, like cannabis, does not seem to be addictive. Nevertheless, it has wound itself deeply into regional culture. Communal qat chews start after lunch, the main meal of the day. Whether at home, at the office or in the shop, everyone brings their own. Merchants chew qat in their shops or stalls. Office underlings are sent out for it, as they are for sandwiches in London or New York. Bus drivers and their passengers chew the leaves in buses; labourers chew at their building sites. Chewing qat is a way of life, and central to hospitality and all social occasions.

Two bundles of qat (*Catha edulis*) prepared for market; the leaves swiftly lose their effect and need to be chewed as soon after harvesting as possible. When bought, these would have been at most two or three days old.

The fresh shoots are brought from mountain terraces generally over 3,000 feet in elevation. The qat that reaches the Hadramout, for example, starts from the Highlands, passes through the ancient cities of Ma'rib and Shabwa, and skirts the Empty Quarter before arriving at Al Qatn. Its popularity is so high that it is having a major effect on the agriculture of the Yemen. Small rises in living standard hugely increase the demand for the foliage. Larger and larger areas of the best farmland have become devoted to this most profitable crop. Though the tree will prosper on dry land, it grows most abundantly if well watered, so it is watered from new boreholes. Consequently the water table is lowering, and natural springs, once part of the ancient and highly developed irrigation system of Arabia Felix, are drying up. Farmers need to grow fewer of the traditional labour-intensive crops to survive. As a result, they rear less livestock and less manure is being put on the land. Qat is already illegal in neighbouring Saudi Arabia, and there are calls for it to be proscribed elsewhere. Yet the farmers need it. The populace want it. Confusion.

Another plant produces similar difficulties, though to an infinitely more intense degree. It, too, allays hunger and fatigue, and acts as a general stimulant, but it contains a powerfully addictive substance. Late in the nineteenth century, it was about to be integrated into the mainstream commerce of America, and indeed the world, when its dangerous dark side was discovered. It is still causing havoc where it is consumed, and havoc where it is grown in South America.

As with qat and cannabis, the place of origin of the coca shrub (*Erythroxylum coca*) is not known. There are several similar wild species growing in the lowlands of eastern South America, Brazil and the West Indies. Coca occupies an ambiguous position among these, and may be a complex hybrid between several of them. The Colombian coca plant is slightly different from the sort grown in Peru and Bolivia and may represent a different genetic mix of the basic species. Line drawings on pottery found in north-western South America show evidence that coca chewing was part of the local culture before the rise of the Inca Empire, perhaps as early as 3000 BC. Coca leaves have been found in Peruvian burial chambers of 2500 BC. The plant was thought to be a gift of the gods, to be used during religious rituals, burials and perhaps on other special occasions. Once the Incas took over the region, the coca leaf was at first officially reserved for Inca royalty. By the time of the Spanish invasion, its use was much more widespread. The new conquerors tried to ban it. In 1551, the Bishop of Cuzco outlawed coca use on pain of death because it was 'an evil agent of the Devil'. The noted sixteenth-century orthodox Catholic artist Don Diego de Robles declared that 'coca is a plant that the devil invented for the total destruction of the natives'. However, the invaders soon discovered that without coca, their enslaved Indians could neither work the fields nor mine gold. Soon landowners, even ecclesiastical ones, distributed leaves three or four times a day.

Later, a Jesuit priest lamented that it was not used in Europe in the way that tea and coffee already were, writing, 'it is melancholy to reflect that the poor of Europe cannot obtain this preservative against hunger and thirst; that our working people are not supported by this strengthening plant in their long continued labours.' Even the English poet Abraham Cowley (1618–67), Queen Henrietta Maria's chief doctor and one of the first members of the Royal Society, cheerfully wrote of it:

Endowed with leaves of wondrous nourishment
Whose juice succ'd in, and to the stomach ta'en
Long hunger and long labour can sustain.

The city of Cuzco, drawn *c.* 1520, with the cathedral firmly planted on the stepped remains of an Inca temple. Ecclesiastical attitudes toward the use of coca leaf changed once its prohibition so affected the productivity of the Indian workers that the economy suffered.

In 1814, an editorial in the *Gentleman's Magazine* urged researchers to begin experimentation so that coca could be used as 'a substitute for food so that people could live a month, now and then, without eating'. By the early nineteenth century, with the Industrial Revolution in full swing and men, women and children in the mills twelve hours a day, six days a week, industrialists began to wonder about using it. It seemed possible that the 'herb' might be used to make workers even more productive. In the 1830s, a German traveller in South America on a hunting expedition at 14,000 feet up in the Andes found that coca was very effective when he took an infusion of the leaves of the plant: 'I could then during the whole day climb the heights, and follow the swift-footed wild animals.'

The active ingredient of coca, cocaine, was first isolated in 1859, by Albert Niemann at the University of Göttingen. It was easy to remove from the leaves using an ether extraction. When cocaine reached North America and Europe, what had once been the divine plant of the Incas became, and remains, a remarkable scourge. At first, it was thought to be an elixir of life. Everyone went wild with enthusiasm. In 1883 a German army doctor tried out the drug on soldiers to see if it did the same for them as the leaves did for the natives of Peru. It did. Cocaine, Dr Theodor Aschenbrandt was able to report, greatly increased their energy and endurance. The report attracted the attention of the psychoanalyst Sigmund Freud. He wrote to his fiancée, 'I am procuring some myself, and will try it with cases of heart disease and also of nervous exhaustion, particularly in the miserable condition after withdrawal of morphine.' Once he'd tried it, he described it as magic, and took it freely for depression and indigestion. Only later did he realize its destructive capacity.

The effectiveness, cheapness and addictiveness of the leaf extract appealed to all kinds of capitalist. Slightly shady ones saw it as the essential ingredient in patent medicines. It was used in toothache cures, cigarettes 'guaranteed to lift depression', chocolates, endless digestives. Women could indulge themselves by taking 'Dr. Worden's Female Pills for Female Diseases and Troubles, Peculiar to the Sex and Women's Delicate System'. Ryno's Hay Fever and Catarrh Remedy, useful for when the 'nose is stuffed up, red and sore', was almost pure cocaine. Combined with alcohol, cocaine forms another potent chemical, cocaethylene. Alcoholic beverages including cocaine, notably Vin Mariani, were taken by prime ministers, royalty and even the Pope. Perhaps fortunately for New York, French enthusiast Frédérick-Auguste Bartholdi (1834–1904) remarked that if only he had used Vin Mariani earlier in his life, he would have designed his statue of 'Liberty Enlightening the World' a few hundred metres higher. The similar Vin Vitae was made of port wine and coca leaves. Incidentally, coca leaves are extremely nutritious, with very high levels of many vitamins.

Most famously of all, in 1886, John S. Pemberton created Coca-Cola. It was promoted as 'a valuable brain-tonic and cure for all nervous afflictions' and a temperance drink 'offering

Though she's about to miss the glass, this speedy lady reflects the enthusiasm with which the mix of alcohol and cocaine was greeted worldwide, amongst anxious searchers for health as well as frivolous hedonists. Only the increasing legislation against *Erythroxylum coca* eventually halted its sales.

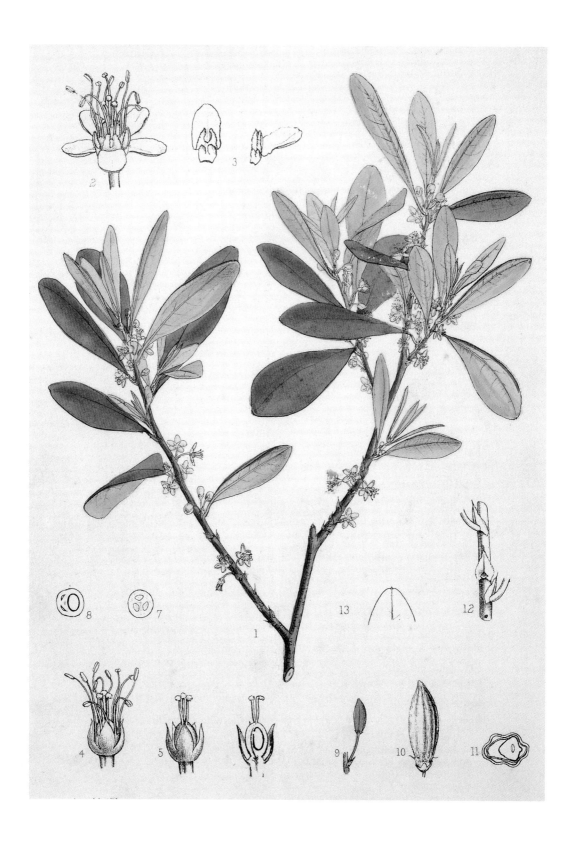

the virtues of coca without the vices of alcohol'. Unsurprisingly, as a typical serving of the new beverage contained around 60mg of cocaine, flavoured with caramel, kola nut, sugar, vanilla, cinnamon and lime, it was a huge success. Purchasers were hooked. After 1903, either the cocaine was removed, or the leaves of a low-cocaine variety of *Erythroxylum* were used to give the drink its special savour.

The range of adherents to cocaine became wider and wider. In the early 1870s, Sir Robert Christison tried it out on medical students in Edinburgh, and was impressed by the results. The chewing of coca leaves, he reported, 'not only removes extreme fatigue, but prevents it'. In France racing cyclists began to take it to increase their powers of endurance, as did the Toronto Lacrosse Club, in Canada, which with its assistance won the title 'Champions of the World'. Sir Arthur Conan Doyle celebrated it with his fictional character Sherlock Holmes, who found that cocaine was 'so transcendentally stimulating and clarifying to the mind that its secondary action is a matter of small moment'. Robert Louis Stevenson wrote *The Strange Case of Dr Jekyll and Mr Hyde* under the South American plant's spell. Ernest Shackleton explored Antarctica propped up by tablets charmingly named 'Forced March', sold over the counter at Harrod's, the London department store, until 1916.

In 1884, the forward-looking pharmaceutical firm Parke-Davis – which said in its advertising copy that cocaine 'could make the coward brave, the silent eloquent, and render the sufferer insensitive to pain' – despatched a scientist, Henry Hurd Rusby (1855–1940), to South America to study coca and look for new sources, and also track down any plants with new medicinal properties that the company could exploit. Rusby arrived in Bolivia in 1885 and spent several months in the Yungas region near Coroico. He later wrote:

The great difference between native coca and other drugs of its class is that its effects do not greatly diminish with continued use. What it does for the Indian at fifteen it does for him at sixty, and a greatly increasing dose is not resorted to … I myself mixed for nearly a year among a million people who use coca daily without ever seeing a single case of chronic cocaism, although this one subject chiefly occupied my attention, and I searched assiduously for information … These people have been described as 'weak, puny, and intellectually little above the beast'. So far as this applies, it is a race peculiarity, and it is the more remarkable that such people should perform daily tasks during long lives which would quickly destroy our finest athletes.

When he took it himself, he found that 'A portion of the results here considered are doubtless due to its anaesthetic effect on the stomach … [hunger] pangs are absolutely allayed by the use of coca leaves. It allays hunger and thirst completely, and thus supports the traveller by removing the chief source of his fatigue.'

Erythroxylum coca is now widely grown in subtropical regions throughout the world, and the sales of cocaine destabilize governments as well as individuals, yet its attractions are sufficient to circumvent all attempts at its eradication.

Yet cocaine is a fierce drug. Crack cocaine, in which cocaine has been cooked up with baking soda, is fiercer still. Like qat, the coca plant is a prized crop for impoverished farmers. The cocaine in it is easily extracted, and very portable. The coca plant, once a useful aid to living in the high mountains, has now become deeply embedded in almost all aspects of life in much of South America, and is an economic engine as important as cannabis and opiates in towns and cities across the globe. It is estimated that worldwide sales are at $92 billion and growing. It is illegal everywhere, banned in the US in 1914, but it was still used at least once a month by 1.7 million Americans in 1998. In the UK it is listed as a Class A substance under the Misuse of Drugs Act (1971). Its illegality, being so widespread, ensures that transportation costs remain high and the mark-up between producer and user can now be colossal.

Coca, like so many other plants in this book, is a Janus plant. Pharmaceutical companies, naturally, continue to hunt for what they hope will be good things without shadow sides. If only they could find plants with coca's powerful hunger-reducing properties, but without the disadvantages of cocaine. They've tried ephedra as a plant that could 'burn' fat. It turned out to burn too much. But primitive man must have often needed plants that lessen the pangs of hunger and thirst. Primitive peoples still do. Wandering over the Kalahari desert in South Africa, hunting bands of San bushmen have to travel long distances with little food or water. So precarious is their survival that some believe their god is a 'trickster' who plays jokes with the land and their fate. One plant was of great use. The natives look out for a strange prickly succulent. A relative of the milkwort (*Asclepias* spp.), it can reach over 6 feet high, and looks rather like an American cactus. The spines, though, are soft, and hunters cut a slice, chew the bitter flesh, and within minutes hunger and thirst evaporate, leaving feelings of strength and alertness. They can then travel for days eating nothing else.

This plant, which the San bushmen call *xhoba*, refers to several species of *Hoodia* that only grow in the region. The bushmen's use of the plant came to the attention of the Council of Scientific and Industrial Research, a South African laboratory partly funded by the government. CSIR isolated the active substance, and patented it as P57. The San, of whom there are still large numbers in South Africa, were not informed or acknowledged. CSIR then licensed P57 to a small British pharmaceutical company called Phytopharm. This company ran trials, and began looking for ways to make semi-synthetic copies of P57. At first CSIR and Phytopharm maintained that the San clan had died out. The San objected loudly, and after legal wrangling an agreement on royalties was reached. The same desert region is also inhabited by another tribe of bushmen, called the Kung. They seem to have made no claim. London's *Observer* newspaper reported Phytopharm's press office claim that P57 would have none of the side effects of many treatments because it was derived from a natural product. At this, Phytopharm's share price rose. However, as many of the stories in this book show, no medicinal plant that works is free from side effects. The CSIR calls itself 'the premier technology and research organisation in Africa committed to innovation, supporting sustainable development and economic growth and creating value for clients, partners and stakeholders'. It eventually issued a press release in which CSIR Bio/Chemtek director, Dr Petro Terblanche said, 'We are proud to be working in domains which require us to enter

Hoodia, a succulent used by the San tribe of the Kalahari desert to reduce hunger pangs in much the same way that coca was used by Andean tribes. It is being taken up with enthusiasm by the pharmaceutical industry, though its mode of action is not clear.

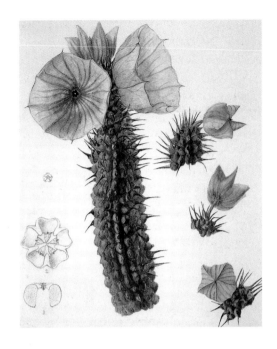

maiden territory in terms of how indigenous knowledge and science interact and how this interaction can best unlock the economic and social benefits inherent in the country's biodiversity.' The San, like so many indigenous peoples, are now mostly settled and demoralized. They, and no doubt the Kung, know of and use other surprising plants that they should urgently be putting under copyright – for instance, they make use of a small plant called *Sceletium*, a relative of the common or garden mesembryanthemum or ice plant, which they mix, when it is available, with ganja, and their shamans get a high.

So many of us like to take something that increases our feelings of ease and well-being: wine, tea, coffee, a cigarette. All these derive from plants that have had a colossal impact on society, and made some people and some places hugely rich, but at huge cost to others. Some, like coca and cannabis, have been proscribed at some stage in their history because they affect the mind. But some of us are beyond such help; we are already in mania, or so wrapped in melancholy that no quick lift can help. To ease the last, huge numbers of plants have been used as antidepressants. The treatment of depression is an area of human nature where the placebo effect is particularly strong. Even modern pharmaceuticals, such as Prozac, evoke it. One psychologist, at the University of Connecticut, believes that the effectiveness of Prozac and similar drugs may be attributed almost entirely to it. An analysis of nineteen clinical trials of antidepressants concluded that the expectation of improvement, not adjustments in brain chemistry, accounted for 75 per cent of the drugs' effectiveness. An earlier study of thirty-nine trials carried out between 1974 and 1995, of depressed patients treated with drugs, psychotherapy, or a combination of both, suggested that 50 per cent of the drug effect was due to the placebo response. Nevertheless, people still search for antidepressants. Ephedra and St John's wort (*Hypericum perforatum*) are currently popular. However, ephedra reacts with substances prescribed for high blood pressure, diabetes or thyroid problems. It also interacts with more conventional antidepressants. Combined with guarana, an inhabitant of the Amazonian rainforest with a huge caffeine content, it can be fatal. St John's wort is also of ambiguous help. It was once thought that hypericin was the active substance, but amounts of this can vary wildly from plant to plant with no apparent change in antidepressant activity. That makes it extremely unlikely that hypericin is an active agent. It is more likely that another substance in the plant, hyperforin, relieves the mind. Alas for the depressed who are also taking digitalis for heart conditions, St John's wort lowers blood levels of digoxin, the most important component of digitalis, by about 25 per cent and that can be dangerous. It also interacts with the cyclosporine used to stop transplanted hearts being rejected by the new owner's immune system. It lowers levels of indinavir and possibly other HIV antiretrovirals. It can also stop oocyte fertilization and alters sperm DNA, making any fertilized embryo liable to deformity.

In the deep past, of course, before all its difficulties were discovered, the poppy plant was the most important 'Destroyer of Grief'. Thomas Sydenham, the seventeenth-century pioneer of English medicine, writes: 'Among the remedies which it has pleased Almighty God to give to man to relieve his sufferings, none is so universal and so efficacious as opium.' Robert Burton's *Anatomy of Melancholy* suggests laudanum for the anxiety-wrecked, made sleepless 'by reason of their continual cares, fears, sorrows, dry brains [which] is a symptom that much crucifies melancholy men'. Failing that, the peony (*Paeonia officinalis*), that garden glory with wine-red, pink or opalescent white flowers, was used. Greek and Roman apothecaries prescribed it, and it is a common find in excavations of medieval monastic sites. Even in the seventeenth century, some folk used the dried roots in amulets, as a prophylactic against madness. Mandrake was used in tiny doses to cure melancholy, though at some risk, and even the humble alyssum of the garden was once called madwort in the belief that it could cure the mad. The humbler thyme was used to calm breakdowns, and became very popular for that purpose in the West Indies when it reached there in the seventeenth century. As a cure for madness, though, neither peony, mandrake or thyme worked.

A plant that does work, to some degree, is the snakeroot or insanity herb (*Rauvolfia serpentina*). It was administered in tiny quantities in Ayurvedic medicine to cure mania and insomnia. It was also used to bring on calm meditative states for those in search of enlightenment. Its use lasted into recent times: Mahatma Gandhi drank a tea made from this herb to relax him after the stress of politics. The West took little notice of it. It may have reached Europe in the sixteenth century, having been brought home, probably from the drug markets of Aleppo, by the German doctor and traveller Leonhart Rauwolff. However, it vanished soon after.

Nothing much else happened with the plant until modern times. In 1943, an Indian physician wrote an article on *Rauvolfia serpentina* that described its sedative effect. In the 1940s, it was widely used to treat over a million Indians for high blood pressure. However, one of its many side effects was to produce severe depression. Nevertheless, Western psychiatrists pricked up their ears. Soon it was being tested on 'mental' patients who were suffering from schizophrenia and paranoia in the hope that it would bring them down to a more normal mental state. An American physician published a positive paper on it in 1952. The plant's alkaloid reserpine rapidly replaced electric shock and lobotomy as treatments for certain types of mental illness. While this may have been of some advantage to the patients, they soon suffered from nightmares, Parkinsonism and gastrointestinal disturbances. Some became suicidally depressed.

The plant's various alkaloids, including rescinnamine and yohimbine as well as reserpine, affect both the central and the peripheral nervous systems. They destroy special parts of the nerve cells in which neurotransmitters are stored. Without these neurotransmitters, the heart rate slows down and the walls of small arteries and capillaries relax. Blood pressure falls. In the central nervous system, the depletion of neurotransmitters causes sedation and the symptoms of schizophrenia lessen. If the plant is then withheld, the nerve cells eventually rebuild their neurotransmitter stores. If not, severe depression follows. Snakeroot is yet another medicinal plant to be approached with great care.

Though we work hard to avoid depression and a sometimes related condition, stress, the plants we have found that seem to alleviate them have, without exception, proved to be fickle friends. If we are caged by the routine of modern society, we become ill. But even those trying to hack a paradise out of a whole new world face enormous stresses. In the 1850s, many of the plants grown by the Shaker community at the Physic Garden at New Lebanon, New York, were used to relieve stress, procure sleep or give some sort of 'high' to counter depression. Even the innocuous lettuce they grew was probably for the sleep-inducing white sap that oozes from the cut stem. Their valerian was for the same purpose.

But let's return to a winter's night in the Paris of 1845. Théophile Gautier is coming down after his dose of hashish. He ends his story:

'Great God!' I cried, struck by a sudden thought. 'If time is no more, when will it be eleven o'clock?' …

'Never!' shouted the resounding voice of Daucus-Carota, thrusting his nose in my face and showing himself in his true colours. 'Never … it will always be nine-fifteen … The hand will remain where it was when time ceased to be, and your corporal punishment will be to observe the motionless hands and return to your seat to start anew, and this until you walk on the bones of your heels.'

'Come now,' said the voyant. 'I can see that it is necessary to exorcise the evil spirits. The thing has gone sour. Let's have a little music. David's Harp will be replaced by the clavichord of Érard.'

And, seating himself upon the stool, he played melodies lively and gay … This appeared to vex greatly the mandrake man, who diminished, flattened, discoloured, and shuddered inarticulately; finally he lost all human form and rolled upon the floor in the shape of a two-rooted salsify. The charm was broken.

'Hallelujah! Time is reborn,' cried childish, joyous voices. 'Go look at the clock now.'

The hand pointed to eleven.

'Monsieur's carriage is waiting below,' said a servant.

The dream had ended. The hachichins each escaped separately to their houses, like the officers after Malbrouck's funeral.

As for myself, I went down that stairway which had caused me such tortures with a light step …

We shall stop him here. He arrived home safely, to live for another twenty-six years. After a productive but impoverished life, he died in 1872. Baudelaire dedicated *Les Fleurs du mal* to him.

THE MYSTERIES
OF THE GODS

Salvia divinorum, a subtropical species of sage, was described as a new species in 1962. It became a 'drug scare' in 2001, and was already part way to becoming illegal in Australia by 31 May of that year. In the United States, much closer to where it grows, it is still legal. Drug enforcement officials say they are monitoring it closely.

Though it has only recently attracted much attention, *Salvia divinorum* is, of course, a plant with an ancient story. It comes from the Oaxaca region of Mexico, an area to which Christianity came late. It is a medicinal plant, used for several centuries by the Cuicatec and Mazatec Indians for curing a range of ailments. The plant helps 'patients' defecate and urinate; however, it stops diarrhoea. It stops headache and relieves rheumatism. It is given to the sick, old or dying to rejuvenate them. Convalescents are given doses of 'la Maria'. It also stops a disease called *panzon de barrego*, or swollen belly. This 'evil eye' complaint is inflicted by a curse from a *brujo*, or evil sorcerer. The swelling is caused by a mysterious 'stone' that has been put inside the sufferer. Only a shaman can remove it. For all these diseases, physical and mental, the plant is administered in a way not so far considered in this book: it is mostly taken only by the shaman-doctor, not the patient. The plant is psychoactive and, under its influence, the shamans use it to divine both the cause of the illness, and the means of its cure. The plant can also answer the sort of questions that many even in the developed world still go to fortune tellers to discuss. What makes the plant a drug scare is that, using higher doses still, the plant gives the shaman direct access to the numinous world – in the case of *Salvia divinorum*, that occupied by the Virgin Mary and St Peter. In other words, it induces hallucinations.

It is almost sterile, and its Mexican, though increasingly worldwide, distribution is achieved by taking cuttings and rooting them. It doesn't seem to occur naturally in the wild, and may be a hybrid between as yet unknown species. As with many subtropical sages, it is a handsome plant, with tall flower spikes and the hook-shaped flowers of palest blue that appear from dark purple-red bracts. It is part of a genus that is widely used medicinally, and

Health and the supernatural have been closely associated in many cultures. Here, in a mural by Mexican artist Diego Rivera (1886–1957), Indian shamans prepare plants for their own divinatory journey to the gods as well as for consumption by the little boy.

in the kitchen, throughout the world. Indeed, the genus name itself comes from the Latin *salvare*, to save. The Middle English name for sage was save. Chaucer, in *The Knight's Tale*, mentions the widely grown *Salvia officinalis* as a cure for wounds and broken limbs. This sage and the annual clary sage (*S. viridis*) have had a long history of use in the treatment of numerous maladies. In China, *S. miltiorhiza*, or danshen, is one of the five astral remedies in the Chinese system of medicine, so important that it shares the same rank as ginseng. It is still in extensive use. Back in Mexico, native Mexicans use nine species medicinally. However, only *S. divinorum* seems to be a hallucinogen.

Amongst Mexican shamans, it is the mildest of the plants used in divination and communication with the saints. It is, though, very important. In shamanic training, the future healer takes *Salvia divinorum* first, in order to learn not only how to heal but also how to identify and use medicinal plants. The 'spirit' of the plant shows the hallucinating trainee a special tree in heaven composed of branches of all the various types of medicinal plant. It also introduces the trainee to God and the saints who give further instruction on how to use each branch. The apprentice shaman – who has been following a ritual diet for sixteen days, during which he or she may not eat foods flavoured with garlic or chilli peppers, and has avoided sex and alcohol – undertakes these 'trips' under the supervision of a master who is not under the plant's influence.

One Mazatec shaman, Maria Sabina (1894–1985), made famous by the writings of many contemporary ethnobotanists, describes in her autobiography (quoted in *Maria Sabina: Her Life and Chants*, written by Alvaro Estrada in 1981) how she used sage to divine what was wrong with a client. She preferred to use some of her more powerfully hallucinogenic repertoire, but when these were not available or in season, *Salvia divinorum* made a good substitute. From her writings, it is clear that Mazatec healing and religion are united in a manner common to traditional cultures. Though the use of *S. divinorum* is pre-Columbian, and was therefore used before Christianity reached the continent, it has become incorporated into the Mazatecs' conceptions of God and the saints, all of whom are seen as the first healers. Of these, St Peter, or San Pedro, is the most influential, and his name is associated with a number of potent drug plants – and with one of the most widespread: tobacco (*Nicotiana* spp.), with which he is supposed to have cured a sick infant Jesus.

For almost all Mesoamerican Indians, pre-Columbian and contemporary, tobacco was and is the most important curing tool in their pharmacopoeia, though not the one most used for contact with the spirit world. It is administered as a snuff, and called 'San Pedro powder', the most efficacious being that ground on 29 June, the saint's day. Alternatively, it is worn in charms and amulets as protection against various 'diseases' and witchcraft. It is also used in ritual cleansings, often combined with other plants such as basil (*Ocimum* spp.) or marijuana (*Cannabis sativa*) and eggs. Tobacco snuff is only one of several sorts used in the Americas.

From many accounts of the seances of Mexican shamans, we know that the leaves of tobacco are gathered by *curanderos* (shamans or healers), who go up into the mountains and harvest them after a session of kneeling and prayer. The foliage is rubbed hard between the

Eighteenth-century snuff takers, caricatured by Thomas Rowlandson (1756–1827). Here, tobacco is being snorted, but in the Americas a large number of plants are used in the same way, their use only sometimes restricted to the shaman.

hands to bruise the leaves before making a tea of them. Around midnight the *curandero*, the patient and another person, who will not drink the tea, go to a dark quiet place, perhaps a distant hut, where the seance won't be interrupted. If the 'patient' is not taking tobacco, but using the tea of *Salvia divinorum* leaves, or ska Maria, he or she begins to go into a semi-delirious trance in about fifteen minutes. As the patient begins to talk, the *curandero*, or *curandera* if a woman, makes a diagnosis. The shaman then closes the seance by bathing the patient in more of the tea. The fresh leaves keep their effect for about a week after collection, though the tea remains potent for only twenty-four hours. After the seance, the used leaves are secretly buried. One of the accounts describes a seance:

> *The curandero picked up a glass of the Maria and began an oration. The Holy Trinity, Saint Peter, the Virgin Mary and other Saints were called on to watch over the participants and teach the visitors the ways of curing. After a final benediction, the potions were drunk and the light was turned out. About 15 min after ingesting the infusion [one of the observers] began to see subtle visions, constricted like columns of smoke in the total darkness. It made no difference whether his eyes were opened or closed. Deciding to speak out, he saw a light which disappeared as he began to describe it. The images increased in intensity. He saw a mountain made of ice, as though he were at the base of a cliff formed from large ice columns. The vision slowly changed into Cerro Rabón, a nearby mountain intimately associated with Mazatec legends … About 23:00 h the flow of images changed into lights of various shades of blue, indigos and purples, scattered as if in a spatial vacuum. Depending on his perspective, he was either traveling through them or else they were being projected toward him. He saw a cross being encircled by a light and a mantle.*

Later, the shaman himself spoke, and even some of the observers who hadn't been having visions began to hallucinate, often seeing images associated with the shaman, or subjects he talked of. Salvia makes the taker very suggestible. Various sensations are reported by those who have taken the tea, all familiar from the witch literature of seventeenth-century Europe:

flying or floating, travelling through 'space', twisting and spinning, heaviness or lightness of the body and 'soreness'. Physical effects also accompany the experience, and if takers try to move, they do so as if very drunk.

The flying sensations are pharmacologically as well as physically surprising, for the plant contains none of the alkaloids present in the old witch plants. Indeed, although its effects have alarmed some and prompted suggestions that it should be illegal, it doesn't seem as if the plant will ever offer much threat to the stability of society. There is little potential for abuse and it is not remotely addictive. More importantly, it may well really be a medicinal plant, for some pharmacologists think that its active substance, salvinorin A, may play a role in the development of new psychotherapeutic drugs, as it seems to affect brain mechanisms in an entirely novel way. It might conceivably give rise to new drugs to help with diseases such as schizophrenia, dementia and bipolar disorders which are characterized by perceptual distortions.

Salvia divinorum seems to be a plant only of the Mazatec and adjoining tribes. However, in their use of another divining plant, the Mazatecs ally themselves with many other tribes of Central America. Ololiuqui, the Aztec name for various sorts of morning glory or bindweed, has been used since prehispanic times and is still popular amongst Zapotecs, Chinantecs, Mazatecs, Mixtecs and other tribes who live in the remote mountains of southern Mexico and northern Guatemala. It has also recently come into fashion amongst experimental Westerners. For the shamans, the most important species embraced by the name ololiuqui is *Rivea corymbosa* (sometimes called *Ipomoea sidaefolia*); the most important for contemporary explorers of the entheogen flora is the wonderful climbing annual of gardens, *Ipomoea purpurea*. The word 'entheogen' was created in 1979, and means 'generating the divine within'. It was intended to replace the now rather judgment-laden misnomer 'hallucinogen', and the culturally freighted term 'psychedelic'. It covers any substance, plant or otherwise, taken by shamans or their followers, in pursuit of the numinous. The numinous can, of course, be merely a revelation in itself, or may impart healing information or knowledge of future events.

There are many colour forms of *Ipomoea purpurea*. Though there are still numerous Aztec names for both plants, *Rivea corymbosa* has been Christianized as *flor de la Virgen*, *semilla de la Virgen*, *hierba Maria* and so on. However, long before Christianity arrived in Mexico the plant was associated with a goddess; in Teotihuacán, a temple frieze shows a mother goddess and attendants entwined with what was called the snake vine. *R. corymbosa* seeds, pale brown, are sometimes thought of as female. The almost black and flattened seeds of *Ipomoea purpurea*, the *badoh negro*, are thought of as male.

Ipomoea purpurea, a fast-growing weed, sumptuously flowered. The hallucinogenic seeds were widely used by Aztec shamans and priests in ancient times. More recently, the plant has been taken up in the West. Many other species of *Ipomoea* have similar properties. This plate is from Curtis's *Botanical Magazine* (1790).

One of the first descriptions of ololiuqui was by Dr Francisco Hernandez (1515–87), a Spanish physician. Between 1570 and 1575, he explored the plants and animals of Mexico for his king, Philip II, and his famous *Rerum medicarum Novae Hispaniae thesaurus, seu plantarum, animalium, mineralium mexicanorum historia* appeared in Rome in 1651. Roughly translated, his description of the plant in this work reads: 'Ololiuqui, which some call coaxihuitl, or snake-plant, is a climbing herb with thin, green, cordate leaves, slender, green round stems, and long white flowers.' He had seen the shamans eating seeds and, during the hallucinations that followed, seen them as they received messages from their gods. In fact, the plant itself was a god, though it was also a medicinal plant that cured conditions as minor as flatulence and headaches and as nasty as venereal diseases and tumours. Ololiuqui is still used by Mexican soothsayers. Seeds are crushed into alcoholic liquids such as pulque, and then consumed by the shaman and sometimes the client too. When the shaman is under the plant's influence, the plant spirit can talk through the shaman to his client. If the client has taken it too, the spirit asks questions that can reveal the seat of the client's illness, and then goes on to suggest a cure.

Both ololiuqui plants, *Rivea corymbosa* and *Ipomoea purpurea*, contain a whole raft of alkaloids, several of which are very similar to the LSD (lysergic acid diethylamide) found in ergot, the fungus that parasitizes the ears of grain crops such as millet, wheat and rye. The discovery of these alkaloids in morning glory was a considerable surprise. Some pharmacologists wonder if the genes for synthesizing these substances may have moved from ergot to the morning glory and its relatives through viral transfer at some distant stage in the past.

The effects of ergot were well known in the ancient world. They were described by the Assyrians in 600 BC. That the effects of consuming the fungus in bread or flour, called ergotism, cause abortions as well as wild hallucinations was noted by the Parsees in 350 BC. Indeed, ergot has been suggested as the hallucinogen used in the mysteries of Eleusis, though there are several other candidates for that particular ritual. The Greeks certainly had problems with ergot. The main crop that became infected was rye, and it happened so often that rye was put under taboo. When the crop began to be widely cultivated in Europe in the Dark Ages, epidemics of ergotism started. Whole communities would suddenly develop wild convulsions and hallucinations. Their arms and legs felt as if they were burning. Dreadful ulcers could occur. Sometimes where nerves died completely, whole limbs could turn gangrenous. The condition was called St Anthony's fire, after the renowned temptations of that saint (d. AD 356). The real cause of the medical condition was not discovered until 1676.

With the modern realization that genetic material can sometimes leap across huge barriers between vastly different organisms, some scientists have wondered if morning glories have incorporated some material from the ergot fungus. Ergot in bread caused the illness called St Anthony's fire, where hallucinations were sometimes followed by the formation of necrotic sores, even gangrene. Here, in the Isenheim altarpiece by Matthias Grünewald, St Anthony is tormented by apparitions, and a sore-covered sufferer has fallen near by.

Other close relatives of *Ipomoea purpurea* also contain exotic substances. The strange climbing 'woodrose' (*Argyreia* spp.), another member of the *Convolvulaceae* or convolvulus family with woody fruits, often sold as a curiosity in seed catalogues, contains LSD-like compounds that shamans avoid, thinking that they might become 'crazy'. In the seventeenth century, the medicinal plant called jalapa was often the dried root of a number of species of ipomoea. It was employed as an emetic, but at least some of the roots were hallucinogenic too. The sweet potato (*Ipomoea batatas*) is currently being studied. Its leaves, together with a number of subtopical plants such as sesame (*Sesamum orientale*, syn. *S. indicum*), castor oil plant (*Ricinus communis*) and *Virola sebifera*, have substances that may control leaf-cutting ants, either attacking the insects directly or reacting in an as yet unknown way with the symbiotic fungi that grow upon the leaf fragments they assemble in their food larders.

The LSD components in the morning glory have been much studied. In animals, they have variable effects. Frogs and mice go into narcosis. Rabbits' hair stands on end, while their blood temperature rises and restlessness increases. In humans, the effect is variable too. Some exhibit only minor effects. Others get the full shamanic 'trip', though some users say it is much less pleasant than the same journey induced by synthetic LSD or natural mescal buttons.

The most famous of the American hallucinogens or entheogens, which contains alkaloids closely related in structure to LSD, has no leaves. Mescal, or peyote (or peyotl), is a strange little cactus, native of the Chihuahuan desert. Until recently, it was harvested each year by the million. Most of the plant is below ground, the juicy tap root topped above the ground by a blue-grey cap, which is marked by a few wavy grooves, and tufts of grey or yellowish

RIGHT *Ipomoea batatas*, the sweet potato, was a great delicacy in sixteenth-century Europe. Like all its close relatives, its leaves have chemicals to control leaf-cutting ants which in many other species seem associated with entheogenic (or hallucinogenic) properties. The illustration shows some of the different flesh and skin colours of the roots.

BELOW The cactus *Lophophora williamsii*, the aerial part shown here being at the top of a large fleshy root; sliced and sometimes dried, it furnishes peyote (peyotl), consumed by a number of native American tribes both as part of religious ritual, and as a medicinal 'tonic'.

fibres. The tufts become denser towards the heart of the plant, and can be decorated with undistinguished whiteish flowers. The cap, or 'button', can be sliced off, leaving the rootstock to grow a cluster of new ones. The buttons are consumed either fresh or dried, and are chopped up and infused like tea. Like so many plants in this book, the local tribes regard *Lophophora williamsii* as a medicinal panacea. It is also, like coca, a plant that allows a man to continue intense physical activity for days without sustenance. Though it was once used mostly by the Cora, Huichol, Seri and Tarahumara cultures, what makes it famous today is that it is an exceptionally powerful entheogen. So many people are now enthusiastic about it, even though its use is illegal in many countries, that the once immense acreages of the 'peyote gardens' of Mexico and Texas have been cleared of plants. It has given rise to great religious and splendidly tangled legal difficulties.

Once the New World was discovered, it didn't take long for the plant to meet the disapproval of the invaders. Francisco Hernandez noted that, after taking it, certain shamans 'can foresee and predict anything; for instance, whether enemies are going to attack them the following day? Whether they will continue in favourable circumstances? Who has stolen household goods? And other things of this sort.' Catholic churchmen wanted to suppress anything so close to what they saw as witchcraft, and in 1620, peyotl was formally denounced by the Spanish:

We, the Inquisitors against heretical perversity and apostasy, by virtue of apostolic authority declare, inasmuch as the herb or root called peyotl has been introduced into these provinces for the purposes of detecting thefts, of divining other happenings, and of foretelling future events, it is an act of superstition, to be condemned as opposed to the purity and integrity of our holy Catholic faith. The fantasies suggest intervention of the devil, the real authority of this vice.

It was an ancient and important part of Indian culture. Bernardino de Sahagún (?1500–90), one of the few Spanish missionaries who learned the local language, was fascinated by the buildings, culture and religion of the Aztec peoples, and detested the methods of Spanish colonization. He made immense researches, which were withheld by the Church until their final publication in Mexico in 1829. In these, he estimated from Indian chronology that peyote had been known to the Chichimeca and Toltec for nearly two thousand years before the arrival of the Europeans. Modern studies of the Tarahumara rock carvings and of Texas rock shelters put peyote's use even earlier.

Little is known of its early ritual use. Today, the Tarahumara tribe, who worship various cacti as gods, only approach the plant with uncovered heads. When they want to harvest one, they make a ritual cleansing of themselves with copal incense. Then, with profound respect, they dig up the god entire. Only men do this; women and children are kept well away from such a dangerous power. Even some Christian Indians regard the cactus god as coequal with Yahweh, and make the sign of the cross in the cactus's presence. Indeed, the plant is surrounded with stories not dissimilar to those wrapped around the European and Egyptian mandrake, a plant almost certainly used in early shamanic rituals of the Old World.

In contemporary Indian peyotl rituals, music is made by rubbing sticks against one another to produce a rasping noise, to which the men and women dance a fantastic and picturesque dance. Sometimes the dance is by the women only, dressed in white petticoats and tunics, and they spin in front of those who are under the influence of the god. The rite, like so many parties in the West, usually takes place on Saturday night. The men sit in a circle around a large campfire. After prayers, the leader hands each man four peyote buttons, which he slowly chews and swallows. The ration for the night is about ten or twelve buttons per man, equivalent to several good-sized plants. Throughout the night the men sit quietly around the fire in a state of reverie, amidst the songs and drumbeats

A hummingbird sips from the headdress of paper and feathers worn by Quetzalcoatl, the Toltec and Aztec god.

of the attendants. As dawn breaks, the colourful visions begin to abate, and by midday the effects have vanished almost entirely. It is time for the participants to attend the service of a more conventional church. If the Saturday night ceremony is attended by someone unused to the effects of the plant, the neophyte is protected against attacks from sorcerers while his or her soul is travelling outside its body. The attendants light a ring of candles around the entranced person. In one of the most extraordinary of contemporary accounts, the female shaman of the seance 'then took a sizeable whole plant, sliced off at the bottom, lifted her magnificently embroidered skirt made specially for the occasion and rubbed the moist end on her legs' – an act reminiscent of the reported use of the mandrake and henbane 'pomade' by the witches' covens of sixteenth- and seventeenth-century Europe.

Naturally, the plant fascinated Westerners. An early account of its effects, entitled 'Mescal: a New Artificial Paradise' and written by Havelock Ellis for the January 1898 issue of *The Contemporary Review*, created a stir:

It has been known for some years that the Kiowa Indians of New Mexico are accustomed to eat, in their religious ceremonies, a certain cactus called Anhalonium Lewinii [an outdated Latin synonym], or mescal button ... it has every claim to rank with hasheesh and the other famous drugs which have procured for men the joys of an artificial paradise. Upon the Kiowa Indians, who first discovered its rare and potent virtues, it has had so strong a fascination that the missionaries among these Indians, finding here a rival to Christianity not yielding to moral suasion, have appealed to the secular arm, and the buying and selling of the drug has been prohibited by Government under severe penalties ... Yet the use of mescal prevails ... has indeed spread, and ... may be said to be to-day the chief religion of all the tribes of the southern plains of the United States ...

On Good Friday I found myself entirely alone in the quiet rooms in the Temple which I occupy when in London, and judged the occasion a fitting one for a personal experiment ... Nothing much happened for a good while, then the visions started [of] thick, glorious fields of jewels, solitary or clustered, sometimes brilliant and sparkling, sometimes with a dull rich glow. Then they would spring up into flower-like shapes beneath my gaze, and then seem to turn into gorgeous butterfly forms or endless folds of glistening, iridescent, fibrous wings of wonderful insects; while sometimes I seemed to be gazing into a vast hollow revolving vessel, on whose polished concave mother-of-pearl surface the hues were swiftly changing. I was surprised, not only by the enormous profusion of the imagery presented to my gaze, but still more by its variety. Perpetually some totally new kind of effect would appear in the field of vision; sometimes there was swift movement, sometimes dull, sombre richness of colour, sometimes glitter and sparkle, once a startling rain of gold, which seemed to approach me. Most usually there was a combination of rich, sober colour, with jewel-like points of brilliant hue ... Sometimes all the different varieties of one colour, as of red, with scarlets, crimsons, pinks, would spring up together, or in quick succession. But in spite of this immense profusion, there was always a certain parsimony and aesthetic value in the colours presented. They were usually associated with form, and never appeared in large masses, or if so, the tone was very delicate. I was further impressed, not only by the brilliance, delicacy, and variety of the colours, but even more by their lovely and various textures – fibrous, woven, polished,

Henry Havelock Ellis (1859–1939), the English psychologist who specialized in sexuality, travelled extensively in the Americas. He became interested in the Amerindian use of plants in the ambiguous area between the spirit world and individual health, lost in Europe for a thousand years.

glowing, dull, veined, semi-transparent – the glowing effects, as of jewels, and the fibrous, as of insects' wings, being perhaps the most prevalent. Although the effects were novel, it frequently happened, as I have already mentioned, that they vaguely recalled known objects. Thus, once the objects presented to me seemed to be made of exquisite porcelain, again they were like elaborate sweetmeats, again of a somewhat Maori style of architecture; and the background of the pictures frequently recalled, both in form and tone, the delicate architectural effects as of lace carved in wood, which we associated with the mouchrabieh work of Cairo. But always the visions grew and changed without any reference to the characteristics of those real objects of which they vaguely reminded me.

He was, of course, entranced. Buying more buttons, he tried them on various artist friends, assuming that their creative natures would engender good visions. One wrote: 'If it should ever chance that the consumption of mescal becomes a habit, the favourite poet of the mescal drinker will certainly be Wordsworth. Not only the general attitude of Wordsworth, but many of his most memorable poems and phrases can not – one is almost tempted to say – be appreciated in their full significance by one who has never been under the influence of mescal.' In spite of such encomiums, it didn't catch on in Europe in the way that cocaine had; perhaps its paradise was simply too strange. It had more success in North America, where the enterprising pharmaceutical company of Parke Davis sold it for its therapeutic use to a small number of psychiatrists and psychotherapists.

In its homeland, the role of the cactus took a different turn. Its use had survived the eighteenth and nineteenth centuries in remote areas. The Indians of Texas and northern Mexico were too racked by war and political strife for most of their old rituals to survive. Oklahoma was more stable, and ritual use of *Lophophora williamsii* survived in the Indian reservation life of the 1870s. The peyote cult was first officially described in 1891. Only a few years before, in 1886, the first attempts at its suppression had been made, by the Indian agent J. Lee Hall, himself later dismissed for drunkenness. However, the report of 1891 defended the religion's legitimacy. It showed that the ancient rites of lophophora had become inextricably entwined with nineteenth-century Christianity. The rather strange fusion is now called the Native American Church, a name that the devotees claimed in 1918. The community professed 'belief in the Christian religion with the practice of the Peyote Sacrament', and the rest of America forgot about it for the next forty years.

In the 1960s, amidst the general hedonistic muddle of the times, mescal began to achieve a wider admiration, especially after the publication of Aldous Huxley's *The Doors of Perception* in 1954 and *Heaven and Hell* two years later, in both of which he described how it felt to use the plants. The enthusiasm for the plant provoked, in America, an immediate backlash. It became illegal in some states, which provoked outrage from the Native American Church. Arizona courts upheld religious protection for peyote, showing that, firstly, it had bona fide religious use. The courts also pointed out that it had no harmful after effects, was not a narcotic, and was not habit-forming. Most importantly of all, for religious native Americans, it was pointed out that to prohibit peyote effectively prohibited the religion itself. The Native American Church immediately began to attract adherents who were not of Indian extraction. Soon too, a non-Indian Peyote Way Church of God in Texas tried to secure the same rights as the NAC. It failed. In North Dakota, a court acquitted a non-Indian couple of possessing peyote because they were long-standing members of a local congregation of Indian peyotists. Then things got confused. A state law proclaimed that religious users of peyote had to be at least a quarter of native American extraction. A district court found that this had the effect of imposing racial exclusion on membership in the NAC, something 'offensive to the very heart of the First Amendment'.

Oddly, whatever the law did or didn't do with respect to *Lophophora williamsii*, it ignored the large numbers of other cacti with almost equally large contents of mescalin and other potent alkaloids – many as sacred to the native Americans as lophophora. 'High' hunters can consume these with ease and impunity. One of the main other peyote genera is the extraordinary *Ariocarpus*. *A. fissuratus* looks like a fistful of rather rough concrete. Like *Lophophora williamsii*, it has a fleshy subterranean tap root and is used for all manner of disease. The juices are used as a painkiller, rubbed on wounds, snakebites and bruises. It is also supposed to dampen fevers and ease rheumatic pains. Often *A. fissuratus* is mixed with water and boiled for a few minutes to make a strongly intoxicating drink. It was also used as a stimulant for native foot-runners, once an important element in the old Inca administrative and communications system. Some Indians carry dried lumps of its flesh as an amulet for good luck or safety from robbers, as this cactus god calls soldiers to its aid.

Another species, *Ariocarpus retusus*, less forbidding to look at, is a punishment for the Huichol Indians. They believe that those who transgress the Huichol ethical code, or who have not ceremonially purified themselves prior to harvesting lophophora roots, will be irrevocably drawn to eat *A. retusus*, and will suffer from 'terrible psychic agonies'. Presumably it gives a bad trip, for the Indians think it an evil plant that drives people mad, even permanently insane. Nevertheless, it is sometimes used to treat fevers.

In southern and western Mexico, the tuberculate cactus *Coryphantha compacta* is respected and feared as a god, with both soul and emotions, almost as powerful as lophophora. On the island of Guadalupe, the native Indians believe that to harm the plant will make them insane, sick or even dead. It can only be touched by the shaman. It, too, is a powerful medicinal panacea: it is masticated and applied to the body to cure all imaginable ills. It is boiled for use as an internal medicine and the juice is applied externally for lung troubles. The chewed pulp is rubbed on the legs of foot-runners for three days prior to the traditional races and is kept in waiting by the shaman should the runner tire. The plant may also be carried in the runners' belts to make them swift and fearless and to frustrate evil spells cast by their opponents. A runner who offends the cactus will run slower and slower and eventually die.

Even the genus *Opuntia*, the familiar 'prickly pear', of which the fruits of some species are now sold in supermarkets worldwide, has entheogenic species. The shiny red fruits of the turkey or coyote cactus (*O. leptocaulis*), when crushed and mixed with an alcoholic drink, are strongly narcotic. Local Indians will not walk close to plants in fruit, fearing that they will be tempted to eat one, and so become inebriated.

In South America, the most famous of all entheogens is the San Pedro cactus, or cactus of the four winds. This is *Echinopsis pachanoi* (syn. *Trichocereus pachanoi*). Its history as an entheogen dates to at least 1500 BC. Many Andean archaeological sites show artefacts that link it with gods and jaguars, especially during the Chavín, Chimú, Nasca, Salinar and Moche periods. Today it is still used by the *curanderos* of Peru, Bolivia and Ecuador, particularly in rites along the shores of lagoons high in the Andes. They make a sacred drink from it called Cimora, the use of which is often associated with the full moon. Often with the admixture of some additional alkaloids from *Brugmansia* species, drinkers are 'set free from matter' and engage in flight through the cosmic regions.

Even away from the deserts, there are members of the cactus family with exotic effects. Beyond the prickly cactuses of conventional shape, the Cactacea family includes many epiphytic genera, ones that grow on tree branches, getting moisture from mist or the rain that penetrates the forest canopy. There are even reports that the humble epiphytic Christmas cactus of the florist (hybrids of *Zygophyllum* and *Rhipsalis*), can contain hallucinogens, but an unidentified species of *Epiphyllum* is used by the Sharanahua people of the Amazon rainforest in their version of the sacred 'ayahuasca'. This, in various recipes, is the hallucinogenic brew used throughout the Amazon basin. The epiphyllum is so strong that the Sharanahua add only one leaf-like phyllode to the brew.

Between Amazonian rainforests and dry Andean slopes, there are a multitude of intermediate habitats in which grows a genus that links the New World and the Old. It also links the contemporary healer and shaman of the Americas with the lost shamanism of the Near East and Europe. In the Americas, all species of *Datura*, and many of the closely related *Brugmansia*, have long been used by native peoples in ceremonies associated with puberty. The most common species is the ubiquitous Jimson weed or thorn apple (*Datura stramonium*). It gets the first of these names from riots, or at least riotous behaviour, amongst British soldiers at Jamestown in Virginia during the year 1676. They are supposed either to have mistaken datura for an edible plant and 'turn'd fool' with hallucinations that endured for eleven days, or to have been deliberately poisoned by disaffected colonists. The leaves smell so unpleasant that the former supposition seems unlikely. *D. stramonium* is now so widely naturalized in America that it is no longer clear if the plant is native or was brought to the continent by settlers. In any case, many species of *Datura* that are clearly native and grow in the warmer parts of the US are just as packed with the trio of alkaloids that have appeared throughout this book: atropine, hyoscyamine and scopolamine. Different species have differing proportions, and usually some unique alkaloids of their own.

Native Americans know the dangers of datura. Some say the plants are not to be touched because they were the first plants made by the gods. Some use its hallucinogenic properties as a rite of passage for their young men. Some call it the 'plant that makes one crazy'. The three alkaloids will at least dampen the pain. It was also used this way in medieval Europe. Gerard noted that 'the juice of Thornapple, boiled with hog's grease, cureth all inflammations whatsoever, all manner of burnings and scaldings, as well of fire, water, boiling lead, gunpowder, as that which comes by lightning and that in very short time, as myself have found in daily practice, to my great credit and profit'. Native Americans had long ago also discovered, as had doctors in India and Ceylon, that the alkaloids of all species of datura could be used against asthma. Sometimes the whole plant was dried and powdered, often combined with saltpetre, to make a sort of incense, whose fumes were inhaled. Alternatively, leaves were dried and smoked. This was effective, paralysing nerves in the pulmonary branches, so stopping bronchial spasm. It was also dangerous. Rajput mothers were reported to smear their breasts with the juice of datura leaves in order to poison their newly born female infants, though it seems more likely that the mothers were merely trying to soothe sore nipples. South and Central American 'daturas' are often perennial and have large drooping flowers that are pollinated by bats, and so are split off now into the genus *Brugmansia*. Some of the species are so splendid that they are cultivated by admiring gardeners all over the world, still referred to as daturas. Most gardeners simply admire them and their marvellous perfumes; in subtropical South America, their alkaloids sometimes play a part in the ancient ayahuasca recipe.

In the rainforest, where there is the greatest density of plant species on the planet, there are huge numbers of entheogenic plants, some long known to the jungle's human inhabitants but only now coming to the notice of ethnographers, botanists and pharmacologists. One widely used is a tree in the genus *Virola*. It is, if distantly, related to the nutmeg tree known to the Queen of Sheba and Cleopatra, and contains some of the same substances. In Colombia, the natives believe the plant to be the 'semen of the sun'. The concentrated sacred juices are kept in containers picturesquely called the 'penis of the sun'.

Every part of the Jimson weed (*Datura stramonium*) contains large amounts of potentially dangerous alkaloids. The immature fruit shown here swells into a large spiny pod. This species is probably originally European but is now naturalized throughout the Americas.

The inner bark contains juices that make powerful arrow poisons, as well as a powerful cure for fungal skin infections such as ringworm and other skin infections that flourish in the humid tropical rainforests. For the latter use, the sap from the inner bark is spread over infected areas of the skin. Some tribes process the bark further. They strip bark from an entire tree and scrape away the inner bark, then squeeze sap from these scrapings, which they boil down until it turns into a hard resin. The resin is allowed to dry, and combined with extracts of other plants. It is then ground into a snuff, and taken just as Amerindians in Central America took, and take, snuff made from ground tobacco leaves. Shamans of the Tukano tribe use virola snuff to consult the spirit world, especially Viho-mahse, the 'snuff-person', who, from his dwelling in the Milky Way, tends all human affairs and can help with their problems.

Virola snuff was not known until the late 1930s, and then only in a most general way. The first detailed description of its preparation and use among medicine men of Colombian Indians was not until 1954. Now it is known through the whole Orinoco basin of Colombia and Venezuela, along the Rio Negro, and in most of western Brazil. Amongst the Colombian Indians, only the shaman uses the snuff, but in other regions all male members of the group above the age of thirteen or fourteen may take it, sometimes in large amounts and over several days. The best snuff is mixed with equal amounts of a powder prepared from the dried, aromatic leaves of a small plant, *Justicia pectoralis* var. *stenophylla*, a close relative of the shrimp plant, recently a popular house plant in northern sitting rooms. Finally, a third ingredient is added: the ashes of the bark of a beautiful tree called *Elizabetha princeps*. Together, the substances from the three plants interact to give a remarkable intoxication, each taker inhaling about a heaped teaspoon of snuff. Users then commonly fall into a wild sleep, and the 'sober' shaman interprets their cries.

Though tobacco snuff is now in worldwide usage, virola snuff has never entered mainstream consumption, which is not surprising because it is only recently that people and plants of Amazonia have been much explored. However, the rainforest, so rich in species, so much under threat and currently so intensely fashionable, does have a plant that is being much more widely taken up in South America, and which must eventually take the world stage. This is a liana actually called ayahuasca, implying that it is the root form of all ayahuascas. The word is a Quechua Indian term meaning 'vine of the souls'. Its scientific name is *Banisteriopsis caapi*. Ayahuasca also refers to the beverage prepared by boiling or soaking the bark and stems. However, the final mix is as complex as virola snuff, and needs to contain several other plants, notably one with the happy name of *Psychotria viridis*. The entheogenic power of ayahuasca is dependent on the way that the alkaloids of the plants interact. Taken unmixed, none contributes a particularly intense effect. It is not known how or when the Amazonian Indians tried out the almost infinite number of plant combinations that the Amazonian forest allows and found this particular combination. However, ayahuasca is widespread through the numerous tribes throughout the Amazon Basin. Nothing is known of its history. There seem to be no ancient myths; the only stories about it are modern, belonging to its various modern user groups. Some Peruvians say that the knowledge came directly from the 'plant teacher', the spirit that inhabits each species and talks directly to the taker during each seance. In Brazil, ayahuasca use has become fused with Christian and African beliefs to make a strange syncretic cult, the UDV. Its members maintain that the knowledge of mixing hallucinogens comes from 'the first scientist', King Solomon. He taught the trick to an Inca king during a journey, perhaps a shamanic one, to the New World.

UDV stands for União do Vegetal. Originally based in Manaus, the cult now has congregations in most Brazilian cities. The organization has its own medical studies section, which carries out research on the effects of the brew that they call hoasca, vegetal, or simply cha (an Oriental word for tea). Many members of the UDV are physicians or psychiatrists, or have other kinds of medical expertise. They have formed an effective pressure group. Banisteropsis and psychotria are permitted by the Brazilian regulatory authorities, who, having studied the church, decided that there was 'no evidence of social disruption, and the ethical and moral behaviour of church members was exemplary'.

Ayahuasca is becoming widely used by white urban Chileans, who, no longer having any connection to native culture, still report thoroughly shamanistic experiences: separation of self from the body, flight, metamorphosis into an animal, and so on.

South America is not alone in having many entheogenic plants: every continent has its own, and nowadays vogues for them sweep across the globe. One increasingly influential plant from central Africa is *Tabernanthe iboga*. Its alkaloid, ibogaine, has been explored as a one-dose cure for heroin addiction, which is marketed under the name 'endabuse' – a pun on the 'Antabuse' which causes alcoholics to be, or at least feel, sick when they take alcohol. Antabuse doesn't always deter alcoholics. Endabuse also sometimes fails. There is also some possibility that large doses of ibogaine can damage the brain. In the Congo and Gabon, iboga is extracted from the plant's roots. At low doses, iboga is a stimulant. At much higher doses, it is an entheogen used in the rite of visitation of the ancestors. Shamans take iboga to seek, in the usual way, information from the spirit world or to ask advice from the ancestors.

Another ritual plant, this time from Oceania, is also finding devotees worldwide. Kava kava (*Piper methysticum*) seems to have been domesticated about three thousand years ago in Vanuatu, a group of islands in eastern Melanesia. The original plant might have been derived from *P. wichmannii* by continued selection over many centuries. There are nearly two hundred and fifty varieties of kava. Now widely grown all over the Pacific region, it was, and still is, used hopefully in the treatment of rheumatism, menstrual problems, venereal disease, tuberculosis and even leprosy. It was also used to procure abortions. What has given the plant global reach lies in its psychoactive substances, kavalactones. These are soporific and narcotic, making the user feel euphoric and tranquil. On the islands, it was, and is taken as an infusion, best prepared by a lovely young girl 'who chewed and infused the kava … [sitting] cross-legged and bare-breasted on a mat behind the kava bowl, with flowers carefully arranged in her hair and her hips swathed in a grass skirt'. This method appalled nineteenth-century missionaries. The process, if picturesque, also turns out to have been unnecessary; the kava is now merely grated, and seems to be just as efficacious. Island men take it during festivities or ceremonies connected with the ancestors and the gods. It offers a way of gaining access to the spirit world. Shamans, or in Hawaii kahunas, used it for divining such matters as how to predict the sex of an unborn child or the cause of illnesses. In the West, it is used as legal 'high', and as a constituent of many herbal medicines, especially ones used as a remedy for anxiety, tension and restlessness. However, it has recently been shown to cause, over time, severe liver toxicity. But whatever happens to kava and iboga, many more plants await their turn to alarm or intrigue us.

It seems as if the young sophisticated Westerner, largely divorced now from superstition and magic, and either intolerant of or untutored about the gods, feels an overpowering need to discover the miraculousness of the surrounding world. If, for some reason, that world is closed off, the use of entheogens can reveal it – as Aldous Huxley wrote in *The Doors of Perception*:

> *I took my pill at eleven. An hour and a half later, I was sitting in my study, looking intently at a small glass vase. The vase contained only three flowers – a full-blown 'Belle of Portugal' rose, shell pink with a hint at every petal's base of a hotter, flamier hue; a large magenta and cream-coloured carnation; and, pale purple at the end of its broken stalk, the bold heraldic blossom of an iris. Fortuitous and provisional, the little nosegay broke all the rules of traditional good taste. At breakfast that morning I had been struck by the lively dissonance of its colours. But that was no longer the point. I was not looking now at an unusual flower arrangement. I was seeing what Adam had seen on the morning of his creation – the miracle, moment by moment, of naked existence.*

The banal had become ecstatic.

But the drive for some of us is the drive to connect with the deep past, back beyond the Inca, beyond Dioscorides, beyond the Vedas and the books of Shen Nung, even beyond the beginnings of civilization. Some want the time of the gods, or of the plants, such as the 'coca mama' of the Inca, who were gods themselves. They want to stand amongst the gods' shadowy attendants, the archaic figures of the shaman, or even that most ancient of all healing trinities, the three hooded ones of the ancient Indus proto-civilizations. Triad or single, shamans were, and are, part priest, part doctor, part soothsayer, even part police detective. They existed at a very early stage in the interface between mankind and the plant world, and where they still exist, they have Western followers eager to touch their ancient world.

Many Europeans, used to the religious hierarchies of the various forms of Christianity and the widespread taboo on personal ecstasy, are fascinated by these primitive figures, who live half in the world of humans and half amongst the spirits, in touch with 'the other' and with healing and wholeness.

In the West, only shadows now remain, some in religious myths where a mortal, Orpheus perhaps, enters the spirit world to retrieve the soul of someone dead – in Orpheus's case, his wife. In other legends, the novice, entranced, undergoes an episode of mystical death and resurrection. In others still, a god, like Dionysus, is associated with an inebriant, just as *Lophophora williamsii* is both god and plant. Traces of shamanism also remain embedded in fairy tales of witches and witchcraft, even in children's myths such as that of Santa Claus, a flying spirit clothed in the snow-dashed scarlet clothes that are a distant memory of the red-capped mushroom amanita. Shamanism remains too in the plant stories associated with 'herbs' such as mandrake and henbane.

While some modern European scholars, notably Mircea Eliade, have felt that the use of entheogens by shamans represents a debasement of the purity of the shamanic world, in fact almost all contemporary shamans, whether amongst Australian aborigines, the Jivaro Indians of central Ecuador and Peru, or the Yakut tribes of Siberia, use plant substances to enter the spirit world. Their rituals often also employ other means for achieving ecstasy: frenzied drumming, dancing, chanting, sleep deprivation, fasting and so on. Some Westerners eagerly attend courses in how to become a shaman. Ethnobotanists and ethnographers, while doing research and publishing papers, find themselves coming to believe in plant gods, in paranormal powers, in the reality of the spirit world. Young people worldwide ransack the shamanic flora for new 'trips' and record their experiences on the Web. More and more people seem to feel increasing distrust of unaccountable hierarchies, whether religious or political or scientific, or medical. Using the shamans' plants, they go in direct and individual search for the numinous.

But for many of us, there are shamans much nearer to hand. Some philosophers, and again Mircea Eliade is an example, maintain that there are in fact shamans in modern Western society: artists. However, few artists nowadays seem in any way to fill the needs of numen hunters, unless by 'artist' the philosophers have musicians in mind. One place where mindset, setting, the plant world and the search for numinous ecstasy do all come together is more banal than an art gallery. It is clubland.

Night-time, a subterranean doorway haunted by dealers, their satchels packed with the plant or plant-derived substances scattered through this book: cocaine, LSD, heroin, cannabinol, speed and of course non-plant psychoactives such as Ecstasy. Inside, darkness, smoke pierced by shimmering lights. The baseline of the music thumps intoxicatingly into the lungs. The dream begins. Gradually, the banal spaces of the club begin to dissolve. Unoriginal music seems good, good music becomes a marvel and marvellous music swirls the potentiated dancer off into the heavens … Indeed, in some tracks the beat ceases, leaving synthesized strings making music to fly with. The club becomes a place of Eleusinian mystery, part witches' sabbath, part shamanic seance. The shaman is the DJ, elevated, in control, interpreting the mood of the dancers, with his eyes turned heavenwards and arms raised in joy as one track segues effortlessly into the next, raising the rhythmic level. For the most fortunate of the dancers, the numinous is there.

William Blake wrote in 1790 in *The Marriage of Heaven and Hell*: 'If the doors of perception were cleansed everything would appear to man as it is, infinite.' From the overperfumed shop in the little town with the castle to the revellers in the smoky depths of the club, the story here is just a fraction of what there is to tell. The full story of the plants that have

changed our lives would be impossible to capture, encompassing as it does myriad medical systems and medicinal floras – there are half a dozen of these in India alone – and myriad sorts of human beings, over a hundred thousand years.

In its relationships with human beings, the plant kingdom has always locked itself closely into our virtues and our vices. It will continue to do so. For however diversely and strangely plants may have interacted with us, there is plainly no such thing as an evil plant; equally, there is no such thing as a good plant. It is only we ourselves who can be good or evil, or both. We can also be fools. There is therefore no reason to believe that future developments in the interaction between man and plants will be all beneficent. Man's future with plants, like the past, will be filled with contradictions and threats. Plants, whether from rainforest, steppe, back yard or the hands of genetic engineers, will continue to produce commodities we have never dreamed of, new trade routes, new wealth and, no doubt, great dangers too.

Adam and Eve in Paradise, by Jan Brueghel and Peter Paul Rubens, *c.* 1615.
The Tree of Knowledge grows amidst the fecundity and ambiguity of Eden; the snake, here evil, was once a symbol of healing. The apple, when eaten, destined Mankind to re-explore the plant world to this day, discovering both its great treasures and its greatest dangers.

BIBLIOGRAPHY

Bastien, J. W., *Healers of the Andes*, 1988

Bisset, N. G., and Wicht, I. M. (eds), *Herbal Drugs and Phytopharmaceuticals*, 2000

Boyle, W., *Herb Doctors in America*, 1988

British Herbal Pharmacopoeia, British Herbal Medicine Association, 1979

British Pharmacopoeia, British Pharmacopoeial Commission, 1988

Burton, R., *The Anatomy of Melancholy*, 1621

Burton, R. F. (trans.), *The Book of the Thousand Nights and a Night*, 1850

Castiglioni, A., *A History of Medicine*, 1941

Coats, J. R. (ed.), *Insecticide Mode of Action*, 1982

Cockren, A., *Alchemy Rediscovered and Restored*, 1940

Culpeper, N., *Culpeper's English Physician and Compleat Herbal*, 1651 (1791 edition cited)

Dioscorides, *De Materia Medica*, trans. 1516

Eliade, M., *Shamanism*, 1951

Emboden, W., *Narcotic Plants*, 1979

'Ethnobotany and the Search for New Drugs', *Symposium 185*, 1994

Felter, H.W., and Lloyd, J.U., *King's American Dispensatory*, 1898

Fowler, G. (ed.), *Mystic Healers and Medicine Shows*, 1997

Gerard J., *The Herball or Generall Historie of Plantes*, 1597 (1633 edition cited)

Grieve, M., *A Modern Herbal*, 1931 (ed. C. F. Leyel, 1985)

Griggs, B., *Green Pharmacy*, 1981

Harton, B., and Castle, T., *The British Flora Medica*, 1877

Harvey, A. M., *et al.*, *The Principles and Practice of Medicine*, nineteenth ed., 1976

Hitti, P. K., *The Arabs: a Short History*, 1996

Holbrook, S. H., *The Golden Age of Quackery*, 1959

Hughes-Hallett, L., *Cleopatra: Histories, Dreams and Distortions*, 1997

Huxley, A., *The Doors of Perception*, 1954

Inglis, B., *The Forbidden Game*, 1975

Josselyn, J., *An Account of Two Voyages to New-England*, 1674

Keys, J. P., *Chinese Herbs, Botany and Chemistry*, 1976

Le Strange, R., *A History of Herbal Plants*, 1977

Leslie, C., *Asian Medical Systems: A Comparative Study*, 1976

Lewis, W. H., *Medical Botany*, 1977

Mabey, R. (ed.), *The Complete New Herbal*, 1991

Marco Polo, *The Book of Ser Marco Polo, the Venetian*, trans. Henry Yule, 1875

McKenna, T. and D., *The Invisible Landscape*, 1975

Meyer, C., *American Folk Medicine*, 1991

Mills, S. Y., *The Essential Book of Herbal Medicine*, 1993

Moore, R. I., *The Formation of a Persecuting Society*, 1991

Morton, J., *Major Medical Plants*, 1977

The Newgate Calendar, 1745–1821

Porkert, M., *Chinese Medicine as a Scientific System*, 1982

Porter, R., *The Greatest Benefit to Mankind: A Medical History*, 1997

Raso, J., *The Dictionary of Metaphysical Healthcare*, 1996

Schultes, R. E., and Hoffmann, A., *Plants of the Gods: Origins of Hallucinogenic Use*, 1979

Shapiro, A. K., and E., *The Powerful Placebo*, 1997

Singer, C., *The Herbal in Antiquity and its Transmission*, 1927

Stark, R., *The Book of Aphrodisiacs*, 1981

Stuart, M., *Encyclopedia of Herbs and Herbalism*, 1999

Swain, A, *Plants in the Development of Modern Medicine*, 1972

Taylor, N., *Plant Drugs that Changed the World*, 1965

Theophrastus, *Inquiry into Plants*, trans. Sir Arthur Holt, 1916

Thompson, R. C., *The Assyrian Herbal: A Monograph on the Assyrian Vegetable Drugs*, 1924

Unschuld, P., *Medicine in China: A History*, 1986

Vogel, V. J., *American Indian Medicine*, 1970

Wasson, R. G., *Persephone's Quest: Entheogens and the Origins of Religion*, 1988

—, 'A New Mexican Psychotropic Drug from the Mint Family', *Botanical Museum Leaflets*, Harvard University, 20: 77–84, 1962

Watt, J. M., and Breyer-Brandwijk, M. G., *The Medicinal and Poisonous Plants of Southern and Eastern Africa*, 1962

INDEX

AUTHOR'S ACKNOWLEDGMENTS

Though I have used information from many websites during the writing of this book, some in particular have been of very great use. In keeping the sea of misinformation at least at some distance, www.skepdic.com and www.hcre.org have been especially useful; www.chem.ox.ac.uk is packed with useful information and www.rpsgb.org.uk has proved its worth; www.erowdid.org is a mine of fascinating material relevant to the last two chapters.

The Shapiros' book *The Powerful Placebo* put huge amounts of material in its correct place, and the late Roy Porter's exciting *The Greatest Benefit to Mankind: A Medical History* gave this book some of its framework.

I should like to thank Dr Henry Noltie and Dr James Ratter, both of the Royal Botanic Garden, Edinburgh – the first for help on Indian and Himalayan plants, the second for help with those of Amazonia. I should also like to thank Jane Hutcheon, librarian at the same institution.

Lastly, I should like to thank my late father: his traveller's tales, and the extraordinary books on his shelves, made me realize how much more interesting it is to try to see the world clearly than simply to believe too much.

PHOTOGRAPHIC ACKNOWLEDGMENTS

The Publishers have made every effort to contact holders of copyright works. Any copyright holders we have been unable to reach are invited to contact the Publishers so that a full acknowledgment may be given in subsequent editions. For permission to reproduce the images on the following pages and for supplying photographs, the Publishers thank those listed below.

AKG, London: 11, 15 below, 17, 36 (Jean-Louis Nou), 38 below, 50, 59, 71, 75, 80 (Erich Lessing), 84, 85, 104, 108, 110, 111, 150, 159, 167, 186 (Erich Lessing)
The Art Archive: 12 (Topkapi Museum, Istanbul/Dagli Orti), 38 above (Palazzo Ducale Urbino/Dagli Orti), 41 (Bibliothèque Nationale, Paris), 91 (Dagli Orti), 113 (Jan Vinchon Numismatist, Paris), 127 (Archaeological Museum, Florence/Dagli Orti), 137 above (Cherokee Indian Museum, North Carolina/Mireille Vautier), 149 (British Library), 168 (Arquivo Nacional da Torre do Tombo Lisbon/Dagli Orti), 180 (Mireille Vautier), 189 (Navy Historical Service, Vincennes/Dagli Orti)
Jennifer Bahney, Longhairlovers.com: 134
Bridgeman Art Library: 5 (Bibliothèque Nationale, Paris/Archives Charmet), 15 above (The Stapleton Collection), 16 (Musée Condé, Chantilly), 18 (The Stapleton Collection), 22 (© The Barnes Foundation, Merion, Pennsylvania), 47, 60–61, 63 (Bibliothèque Nationale, Paris), 72 (Private Collection), 76, 83 (Private Collection), 94–5 (Bibliothèque Nationale, Paris), 97 (Egyptian National Museum, Cairo), 119 (The Stapleton Collection), 120 (Musée de la Ville de Paris, Musée Carnavalet, Paris), 132–3 (Bibliothèque Nationale, Paris/Archives Charmet), 142 (Biblioteca Medicea-Laurenziana, Florence), 153 (Natural History Museum, London), 154–5 (Bibliothèque Nationale, Paris), 156 (Musée de la Vie Romantique, Paris), 162 (Bibliothèque Nationale, Paris), 164 (The Stapleton Collection), 173 (Private Collection/ Archives Charmet), 183 (Bradford Art Galleries and Museums), 191 (Biblioteca Nazionale Centrale, Florence), 202–3 (Mauritshuis, The Hague)
By permission of the British Library: 79, 129
© Christie's Images Limited 2004: 86–7, 161
Mary Evans Picture Library: 147, 171, 193
Bob Hohertz: 67, 139
Courtesy of Hunt Institute for Botanical Documentation, Carnegie Mellon University, Pittsburgh, PA: 28, 124
Dr Warnar A.W. Moll, Amsterdam: 27, 44
Museum of Garden History: endpapers, 8, 45
Private Collection: 35, 52, 122,
Royal Botanic Gardens, Edinburgh: 1,170
Royal Botanic Gardens, Kew: 2, 25, 31, 54, 74, 89, 125, 140, 177, 185, 188
Royal Horticultural Society, Lindley Library: 49, 57, 93, 100–101, 131, 144, 174, 197

PUBLISHERS' ACKNOWLEDGMENTS

Project editor Anne Askwith
Editorial assistance Serena Dilnot
Picture editor Sue Gladstone
Picture assistant Milena Michalski
Index by Margot Levy
Designed by Anne Wilson
Production Caterina Favaretto